SATELLITE EARTH OBSERVATION

A COMPENDIUM OF CAPABILITIES AND APPLICATIONS FOR DEVELOPMENT

JULY 2026

ASIAN DEVELOPMENT BANK

6 ADB Avenue, Mandaluyong City, 1550 Metro Manila, Philippines
Tel +63 2 8632 4444; Fax +63 2 8636 2444
www.adb.org

ISBN 978-92-9277-730-2 (print); 978-92-9277-731-9 (PDF); 978-92-9277-732-6 (ebook)
Publication Stock No. TCS260092-2
DOI: http://dx.doi.org/10.22617/TCS260092-2

Notes:
In this publication, "CNY" refers to the yuan, "€" refers to the euro, "¥" refers to the Japanese yen, and "$" refers to United States dollars.
ADB recognizes "China" as the People's Republic of China, "Korea" as the Republic of Korea, and "Vietnam" as Viet Nam.
Effective 1 February 2021, ADB placed a temporary hold on sovereign project disbursements and new contracts in Myanmar.

Cover design by Francis Manio.

On the cover: Main photo: Copernicus Sentinel 1 satellite useful for conducting field studies to better understand and support development projects (photo by ESA/EU). Background photos: (i) city skyline, (ii) people doing research in the forest, and (iii) wind turbines (photos by ADB).

CONTENTS

TABLES, FIGURES, AND MAPS

TABLES

FIGURES

MAPS

FOREWORD

New digital technologies continue to develop at an exponential rate and profoundly impact the way we live and work. These technologies offer scalable solutions to help address development challenges, combat the deleterious effects of climate variability, and advance a just transition to clean energy.

International financial institutions and development practitioners face increasing pressure to manage and implement their activities in an environmentally sustainable manner, that is, by avoiding harm to environmental ecosystems, habitats, and biodiversity. As such, they are in constant need of a wide range of accurate contemporary and historical environmental information on local, national, and regional scales. The unavailability of required environmental information causes continual problems for the efficient implementation of development activities. Satellite Earth Observation (EO) is a growing digital technology offering enormous potential to address these sources of problems and help solve development challenges by providing such key environmental information in a globally consistent way. Therefore, the Asian Development Bank (ADB) is committed to supporting EO technologies as part of its goal to enable digital development across Asia and the Pacific.

Over the last decade, ADB operational departments and resident missions have been exploring the potential of environmental information derived from satellite EO imagery and data for development projects. These initial experiences led to valuable insights into the benefits and practical use of EO technology in policymaking, planning, and the full range of ADB's operations.

This technical study, *Satellite Earth Observation: A Compendium of National Space Capabilities and Applications for Development*, has been commissioned by the Digital Sector Office. This report is a companion document to the shorter ADB Working Paper No. 99, *Satellite Earth Observation Assets and Solutions: Supporting Operations in Asia and the Pacific*, published in November 2024, which addresses the main concepts and issues related to the use of satellite EO in support of development operations. This previous working paper broadly covered the main concepts at a high level to more easily disseminate information and stimulate discussion.

This technical report covers the same EO domain in significantly greater detail for ADB and its developing member country users that want to explore the subject further. In particular, the report presents an extensive overview of the EO satellite landscape (EO space segment, EO data access and processing, international initiatives, the current market for EO data and information services, the benefits of using EO information for decision-making, and a summary of potential future trends in the EO sector). In addition, and representing the core of this report, are the detailed and specific examples of EO environmental information services for six thematic use cases taken from current ADB operations. These use cases cover the domains of (i) basic land surface type (cover) mapping and monitoring, (ii) disaster risk management and disaster risk reduction, (iii) climate resilience and proofing, (iv) coastal zone management, (v) forest monitoring and management, and (vi) greenhouse gases monitoring (CO_2, CH_4, N_2O).

I hope that this report will enable development practitioners in ADB and across its developing member countries to better understand and apply satellite EO technologies, thereby fostering our shared vision of a prosperous, inclusive, resilient, and sustainable Asia and the Pacific.

Antonio Garcia Zaballos
Director, Digital Sector Office
Asian Development Bank

ABOUT THE AUTHORS

Paolo Manunta serves as a senior digital technology specialist in the Digital Sector Office at the Asian Development Bank (ADB). He specializes in Earth Observation (EO), online geographic information systems (GIS), and spatial data infrastructure (SDI). In this role, Paolo leads the establishment of geospatial initiatives and has facilitated the cloud-based integration of EO, GIS, and artificial intelligence (AI) across various ADB developing member countries.

Paolo joined ADB as an ESA Resident Support, with his experience in Asia and the Pacific beginning in 2014 through a transport project in Papua New Guinea. He transitioned to a full-time secondment at ADB in 2017 and became a permanent staff member in 2022. Following the completion of his PhD from the University of Alberta in Canada, as data analyst he worked on climate change modeling and developed adaptation pathways for the Provincial Government of Alberta in Canada. He has over 15 years experience in the private sector in Europe.

Stephen Coulson has 30 years of experience in the field of satellite EO and its environmental applications with the ESA based in the European Space Research Institute (ESA/ESRIN), Frascati, Italy. Since 2015, he has managed a set of activities focused on developing initial cooperation with major international development banks, e.g., ADB, the World Bank, the Inter-American Development Bank, the European Investment Bank, the International Fund for Agricultural Development, and the European Bank for Reconstruction and Development, in order to demonstrate the benefits and expand the use of EO-based environmental information in development assistance context. These activities led to the start of a new, dedicated 5-year ESA program in 2020 (Global Development Assistance, or GDA) to implement the required technical developments (e.g., EO information products, user-oriented analytics tools) that best respond to operational needs in development. He retired from ESA in July 2021.

ACKNOWLEDGMENTS

This report was prepared as part of the implementation of the Asian Development Bank (ADB) regional technical assistance, Empowering Developing Member Countries to Use Multispectral Satellite Images and Artificial Intelligence for Land Use and Coastal Planning, co-financed by the Republic of Korea e-Asia and Knowledge Partnership Fund and the High-Level Technology Fund. Paolo Manunta, senior digital technology specialist (Earth Observation), Digital Sector Office, Sectors Department 2 (SD2-DIG), led the development of this technical study with Stephen Coulson as the main consultant. Overall guidance was provided by Antonio Garcia Zaballos, director, SD2-DIG; and Inhyok Kwon, digital technology specialist (digital government), SD2-DIG. In this technical study report, research was conducted to provide the context for accessing publicly available documents and open-source data, including research reports, media articles, and academic papers.

This publication was produced with the support of a team comprising Cyrel San Gabriel as copyeditor, Lawrence Casiraya as proofreader, Alvin Tubio as typesetter and layout artist, and Francis Manio as graphic designer. Genny Mabunga, operations analyst, SD2-DIG and Carmela Fernando-Villamar, senior digital technology officer, SD2-DIG provided valuable administrative support. The peer reviewer of this technical study report was Manzul Hazarika, director, Geoinformatics Center, Asian Institute of Technology, whose feedback and advice were most valuable.

ABBREVIATIONS

AaS	as a service
AI	artificial intelligence
AIS	automatic identification system
API	applications programming interface
APRSAF	Asia-Pacific Regional Space Agency Forum
ARDC	Africa Regional Data Cube
BRIN	National Research and Innovation Agency (Indonesia)
C3S	Copernicus Climate Change Service
CBERS	China–Brazil Earth Resources Satellite
CCI	Climate Change Initiative
CEOS	Committee on Earth Observing Satellites
CGAR	compound annual growth rate
CNES	Centre National d'Etudes Spatiales
DMC	Disaster Monitoring Constellation
DRM	disaster risk management
DRR	disaster risk reduction
DTO	Digital Twin Ocean
EARSC	European Association of Remote Sensing Companies
ECV	Essential Climate Variables
EO	Earth Observation
EO4SD	Earth Observation for Sustainable Development
ERS-1	European Remote Sensing Satellite-1
ESA	European Space Agency
ESDT	earth system digital twins
EUMETSAT	European Organization for the Exploitation of Meteorological Satellites
EUSPA	European Union European Space Progammes Agency
FFO	full, free, and open
GCOS	Global Climate Observing System
GEO	Group on Earth Observations
GHG	greenhouse gas
GNDVI	green normalized difference vegetation index
GNSS	global navigation satellite system
GST	Global Stocktake
IPB	Institute Pertanian Bogor (Indonesia)

IPCC	Intergovernmental Panel on Climate Change
ISF	informal settlement family
ISRO	Indian Space Research Organization
JAXA	Japan Aerospace Exploration Agency
KARI	Korea Aerospace Research Institute
LAPAN	National Institute of Aeronautics and Space (Indonesia)
LiDAR	light detection and ranging
LULC	land use and land cover
NASA	National Aeronautics and Space Administration
NCEI	National Centers for Environmental Information
NOAA	National Oceanic and Atmospheric Administration
PhilSA	Philippine Space Agency
PRC	People's Republic of China
SAR	synthetic aperture radar
SAVI	soil-adjusted vegetation index
SDG	Sustainable Development Goal
SDI	spatial data infrastructure
SEOS	satellite earth observation system
SPAC	special purpose acquisition company
SPOT	Satellite pour l'Observation de la Terre
SSTL	Surrey Satellite Technology Ltd.
TEP	Thematic Exploitation Platform
UN-GGIM	United Nations Committee of Experts on Global Geospatial Information Management
UNFCCC	United Nations Framework Convention on Climate Change
UNITAR	United Nations Institute for Training and Research
UNOOSA	United Nations Office for Outer Space Affairs
UNOSAT	United Nations Satellite Centre
VAS	value-adding services
VC	Venture Capital
VHR	very high resolution

EXECUTIVE SUMMARY

The purpose of this technical report, *Satellite Earth Observation: A Compendium of National Space Capabilities and Applications for Development,* is to provide a more detailed insight into the broader Earth Observation (EO) landscape. This involves the main players and operators in the EO space segment (both government and private sectors), the main processes and actions in turning basic EO data into information services fit for decision-making, a summary of key international EO initiatives, a review of current EO market estimates and growth forecasts, a description of top-down and bottom-up methodologies to estimate the benefits of using EO information in daily work, and a summary of some of the future trends that are taking place in the EO domain.

In addition, this technical report provides detailed and extensive examples of specific EO environmental information services that are currently being used to support ADB development operations (i.e., EO use cases). These information services are grouped into six thematic domains: basic land surface type (cover) mapping and monitoring, disaster risk management (DRM) and disaster risk reduction (DRR), climate resilience and proofing, coastal zone management, forest monitoring and management, and greenhouse gas (GHG) monitoring (carbon dioxide [CO_2], methane [CH_4], and nitrous oxide [N_2O]). An additional use case looks at how EO-based information on the urban environment can help monitor access to safe and affordable housing and basic services.

Overview of the satellite EO landscape. EO satellites are traditionally large (i.e., a few thousand kilograms [kg]), expensive (i.e., a few hundred million euros), and have complex engineering systems (i.e., more than 5–10 years of development period). Developed countries have made substantial investments in satellite EO over the last 50 years. These capabilities contribute to global efforts in environmental monitoring, disaster management, and sustainable natural resource utilization. A leading example of this traditional satellite EO approach is that of Copernicus, the EO program of the European Union (EU), which is the most comprehensive, complete, and capable EO satellite system in the world today. The Copernicus Program is funded, coordinated, and managed by the European Commission, in cooperation with partners such as the European Space Agency (ESA), the European Organization for the Exploitation of Meteorological Satellites (EUMETSAT), and the European Environment Agency. The ESA has more than 30 years of experience in the EO domain in developing countries, taking a lead in Earth science and EO-based applications and informational products. Copernicus consists of a set of dedicated, purpose-built satellites (the Sentinel families), and the so-called "Contributing Missions," which are third-party commercial and public satellites. In 2023, there were eight Sentinel satellites in orbit comprising five different families. Copernicus has been a game changer in EO global data acquisition and access, collecting about 25 terabytes (TB) of data each day (as of late 2023). Copernicus monitors our planet, and through the full, free, and open (FFO) data policy, delivers free data and environmental information services based on satellite EO data, and in situ (non-space) data to all users globally. Since the launch of Sentinel-1A in 2014, the EU has started a process to place a constellation of almost 20 more Sentinel satellites in orbit before 2030. In addition to these large, international initiatives, many countries in Asia and the Pacific have extensive national space programs following this traditional public sector development approach. Some examples include the People's Republic of China's Gaofen and GaoJing series of optical satellites for EO, Fengyun series for meteorology, Haiyang series for oceanography, and Huanjing series for environmental and disaster monitoring; Japan's Advanced Land Observing Satellite (ALOS) series, Global Change Observation Mission – Climate (GCOM) series, and Greenhouse Gases Observing Satellite (GOSAT) series; and India's EOS series (for agriculture, forestry and plantations, and disaster management support), Cartosat series (for urban development and planning, management of rural resources and infrastructure development, and monitoring of coastal land use and land cover), RESOURCESAT series (for natural resources monitoring), and IRS series (for agriculture and land cover).

Alongside this traditional satellite EO approach, a range of exciting, radical, and highly innovative developments have taken place over the last 10 years, known collectively as *NewSpace*. These disruptive technology developments are brought by private companies and small startups entering the space industry and have led to the emergence of so-called small, micro, nano satellites; whereas the satellites are small (i.e., 1–100 kg), affordable (i.e., less than 1 million euros, including launch and operational costs), and built from standard engineering subsystems (i.e., development timeframes of a few months). Small satellites (SmallSats) can range in size from that of a shoebox to a washing machine and weigh as little as 5–10 kg. The NewSpace revolution is being achieved through the use of readily available, commercial, off-the-shelf components for SmallSat electronics and structure (like those mass-produced for cell phones, automotive industry, and recreational drones), thereby keeping cost and delivery times low. These factors have led to completely different satellite missions and business models, with constellations of hundreds of nano satellites (Nanosats) operating together to significantly increase global coverage and frequency of observation of the Earth's surface. The NewSpace trend began and is well advanced in the United States (US), with Europe now playing catch-up. However, Asia and the Pacific is now poised to be the fastest-growing region for this innovative technology sector over the next decade. Many economies in the region have advanced capabilities in the development, production, and operation of microsatellites and CubeSats. They have made significant strides forward in satellite EO, harnessing space-based technologies for a wide range of applications, including disaster management, agriculture, resource monitoring, and environmental protection. Strong capabilities in this sector already exist in Asia and the Pacific, with some examples as follows: the FORMOSAT satellite series in Taipei,China: the DIWATA and MAYA series in the Philippines; the Arirang and Cheollian series in the Republic of Korea; and the BRIN NUSANTARA constellations in Indonesia.

On the ground—and complementing EO satellites in space—are EO data acquisition activities and access; EO data processing, fusion, and analytics; and the provision of both standard and customized EO-derived environmental geo-information products and services for decision-making. These activities constitute the generic concept of the EO information "value-chain." Collaboration between economies and international organizations in the domain of EO has increased data accessibility and sharing, further enhancing the benefits of EO on a global scale. New and innovative geospatial data processing infrastructures and platforms are constantly emerging and are key to realizing the potential of EO, especially given the volume and complexity of data now being produced. With the advent of high-performance cloud computing infrastructures and data analysis techniques, complex environmental monitoring problems are beginning to get solved with EO-based information. Using artificial intelligence (AI) and machine learning techniques can help decipher the exponentially growing volume of EO data, as computers have the capability to identify patterns and correlate vast volumes of data that conventional analytic approaches cannot handle. The ADB Cloud-based Satellite Earth Observation System (Cloud SEOS) is a prototype cloud-based platform currently under deployment in Southeast Asia. It supports the use of machine learning techniques to allow key stakeholders to process free and open satellite data from sensors at different spatial resolutions. The aim of Cloud SEOS is to host AI models and satellite EO-based imagery, resulting in open and cloud-based platforms to facilitate capacity building and to foster uptake in developing member countries lacking EO skills and major investments in local IT infrastructure. Data cubes are multidimensional arrays of stacked, EO images and data aggregated from a variety of sources, but standardized and harmonized to be “analysis ready” (i.e., to be used in further analytics with minimal effort). Data cubes are deployed on the cloud for efficient, on-demand data processing. A digital twin is a virtual model of a physical object. A digital twin works by digitally replicating a physical asset in the virtual environment, including its functionality, features, and behavior. Digital twin Earth developments will help visualize, monitor, and forecast natural and human activity on the planet. Finally, EO value-adding services (VAS) are generated and provided by a vast and highly diverse range of both public sector organizations and rapidly growing private sector companies. The European EO VAS industry is well developed in terms of technical capability due to a long history of public investment, both at the European (EU, ESA programs) and national levels. For developing countries, the level of activity in the EO VAS domain is significantly smaller, with about 241 companies active in the EO industry in 2019. In addition to the VAS and analytics tailored to users' specific needs and providing targeted solutions to specific environmental issues, there is a wide range of global environment informational products available that are more standardized and can be accessed via digital platforms and portals free of charge.

The satellite EO industry sector is highly competitive with many players in the market. In 2022, the EO data market stood at $1.78 billion (2022) and is growing at a 5-year compound annual growth rate (CAGR) of 6%. Forecasts project future growth to reach a total value of more than $2.7 billion by 2032 at a 4% CAGR throughout the decade. As of 2021, EO data were being sold commercially from more than 370 satellites, with this number forecast to grow to more than 1,000 satellites in 2030. The EO VAS and analytics market is valued at $2.86 billion (as of 2022) and demonstrates a 5-year CAGR of 9%. Predictions envision an escalation to $4.9 billion by 2032 at a 6% CAGR. North America is the largest geographic region in terms of EO market share; however, Asia and Pacific is estimated to grow at the highest CAGR over the period of 2023–2027. For developing countries, the data suggests that EO companies based in these countries contributed a total of €82 million to the global EO VAS market in 2019 (2.2% of the global market). However, this relatively low contribution to the global EO market represents a significant opportunity for future growth in the digital economy of these countries.

The benefits attained through using satellite EO environmental information can be estimated via both top-down and bottom-up analyses. Top-down methods include quantitative impact assessments using a thorough methodology to explore how people use improved environmental information to make better decisions. The studies quantify how these decisions impact the defined socioeconomic metrics, such as lives saved or resources conserved. In addition, the European Commission has published several fact sheets that define and demonstrate quantifiable, practical examples of benefits achieved by using the Copernicus Sentinels data. The types of benefits looked at comprise cost reductions, additional revenues, and environmental improvements. Each fact sheet addresses a specific sector, including agriculture, forestry, insurance, ocean monitoring, oil and gas, renewable energies, urban monitoring, and air quality. A good example of a systematic approach and definition of a detailed bottom-up methodology in estimating the benefits of using satellite EO information is the Sentinels Benefits Study (SeBS). The purpose of the SeBS benefits analysis is to understand how a specific EO-derived service is being used by an organization, which in turn is benefiting others and ultimately the society and citizens at large. The bottom-up approach is based on the detailed evaluation of the EO value chain for a specific use case. More than 30 use cases have been investigated leading to a comprehensive and detailed understanding of the benefits that EO can deliver in these domains.

In terms of business models and future trends, the EO domain is currently experiencing drastic changes that will have strong impacts on the EO industry. The EO market is rapidly evolving and attracting a growing number of investors, notably in the US. Key drivers include the rising use of SmallSats deployed in large constellations (hundreds of satellites) providing global coverage and high revisit rates even at very high resolutions of a few meters; the increasing availability of private sector financing from business angels to venture capitalists, financial corporations, or the public markets; the increasing use of innovative technology and appetite for risk-taking in the "fail and rebuild fast and cheap" development approach; and the rapid advance in data analytics and value-added information services through the use of enhanced computing facilities and data processing technologies (platforms). *"Space-as-a-service"* models are emerging, allowing governments to contribute to EO missions by supplying critical payloads, while "outsourcing" the rest of the space segment to commercial companies. *"Satellite-as-a-service"* models are a recent development allowing countries and companies to acquire ready-made satellites directly from industrial manufacturers, thereby ensuring that the assets are fully owned by customers. Finally, *"data-as-a-service," "platform-as-a-service," and "insight-as-a-service"* models complete the EO information delivery stages where rapid evolution is taking place, with go-to-market strategies being employed by commercial EO companies as they scale-up businesses and focus on the impact of EO on customer processes and how it helps them accomplish their goals. Across the broader geospatial industry, the next 5 to 10 years will see significant developments in the maturity and application of already well-established technologies. Among others, AI, sensor technology, and the Internet of Things will drastically change the way in which data is collected, managed, and maintained.

Example use cases of satellite EO in ADB development operations. There is a vast number of sensors and instruments currently being flown on EO satellites that measure an even greater range of physical properties on the Earth's surface. Satellite EO provides a wide range of key thematic types of environmental information that is global, comprehensive, accurate, repeatable, and timely. Some of these main thematic information types include the following:

- Land surface type (cover) mapping, vegetation and biomass index mapping, biodiversity mapping for natural resource monitoring and management.
- Urban areas and asset mapping at high and very high resolution for improved urban planning and management.
- Natural hazard mapping (geo-hazards, hydro-meteorological hazards, and climatological hazards, both historical and event-based) and risk assessment and modeling for disaster risk management and reduction.
- Derivation of a set of climate indicators providing information about climate risks for improved climate resilience and proofing actions. The main EO informational content covers a wide variety of environmental information: for the domains of *Land* (e.g., soil moisture, snow extent, glacier extent, lake level, lake water quality, land cover, land use, land surface temperature, evapotranspiration, drought, fire); *Ocean* (e.g., coastal sea level, water quality, met-ocean conditions); and *Atmosphere* (e.g., precipitation, air temperature, greenhouse gas concentration, air quality).
- Coastal environmental parameters (including Chlorophyll-A concentration in μg, transparency or turbidity in m^{-1}, sediment concentration in μg/l, algal bloom detection and delineation), coastal met-ocean information (sea surface temperature on the Kelvin scale, coastal wave information, and coastal currents), coastal hydrographic information (including bathymetry and the evolution of coastline mapping to monitor erosion and deposition dynamics), and coastal habitat status (including mangroves, coastal mud flats, sea grass beds, and coral reefs) for improved coastal zone management.

- Basic forest type mapping (typically 2–4 classes within deciduous and coniferous forests). For tropical forests, they are typically classified as broad-leaved forest, bamboo, palm forest, mixed forest or forest plantations; forest cover change mapping (afforestation, deforestation and reforestation, e.g., canopy gaps, clear cuts, burnt areas, and regrowth); and mapping of forest biomass and carbon statistics for improved forest management.
- Mapping of the global spatial location, and distribution and concentration of main GHG sources and emissions (i.e., CO_2, CH_4, and N_2O) for management issues related to the atmosphere composition.

Detailed examples of all these EO information services are presented in the context of current ADB development operational activities. These information services are grouped into six thematic domains (or EO use cases): basic land surface type (cover) mapping and monitoring, DRM and DRR, climate resilience and proofing, coastal zone management, forest monitoring and management, and GHG monitoring (CO_2, CH_4, N_2O).

An additional use case demonstrates how specific EO-based urban mapping information can help in understanding access to basic services and adequate, safe, and affordable housing, which could help monitor progress toward SDG target 11.1. This draws on land use and land cover (LULC) mapping of Metro Manila for four points in time (1990, 2000, 2010, and 2014); change detection and analysis of the LULC time series to understand urban dynamics; and data on informal settlements.

Conclusions. The most comprehensive, complete, and capable EO satellite system in the world today is Copernicus, the EO flagship program of the EU. Copernicus is a long-term program started in 2014 to develop, build, and launch a constellation consisting of almost 30 Sentinel satellites in orbit before 2030. However, over the last decade, the satellite EO domain has (and continues to be) transformed through a range of exciting, disruptive, and highly innovative developments known collectively as NewSpace. This emerging model of NewSpace is characterized by the proliferation of private companies and small startups entering the space industry with radically different approaches, technologies, and business models. NewSpace is driven by constellations of many micro and nano satellites, built mainly with commercial components and through simplified production and test processes that rely on low complexity. This new trend began and is well advanced in the US, with Europe now catching up. However, Asia and the Pacific is now poised to be the fastest-growing region for this innovative NewSpace technology sector over the next decade. Satellite EO provides a wide range of key thematic types of environmental information that is global, comprehensive, accurate, repeatable, and timely. Satellite EO is still the only means to provide comprehensive, consistent, and up-to-date imagery of the Earth's global surface and of the changes taking place on that surface, both in time and spatially. Regarding the seven operational priorities of ADB Strategy 2030, EO is already making important and useful contributions to several of these (as identified in the ADB use cases) and has the potential to grow significantly in the future.

INTRODUCTION 1

Over the last decade, the Asian Development Bank (ADB) has been exploring the potential of environmental information derived from satellite Earth Observation (EO) image data to be used in development operations, planning, and policymaking.

This exploration has been taken forward through strategic partnerships with the Japan Aerospace Exploration Agency (JAXA) and the European Space Agency (ESA), both of which have or had resident support in the form of EO experts located in ADB headquarters to assist ADB operations staff, resident missions, and developing member countries in using EO environmental information in their daily work.

This technical study, *Satellite Earth Observation: A Compendium of National Space Capabilities and Applications for Development,* is a companion document to a shorter ADB working paper, *Satellite Earth Observation Assets and Solutions.*[1] This working paper addresses the main concepts and issues related to the use of satellite EO in support of development operations and broadly covers the main concepts, but at a high level only, to more easily disseminate information and stimulate discussion.

This technical report covers the same EO domain in significantly greater detail for ADB and developing member country users that want to explore the subject in further detail. In particular, the report presents an extensive overview of the EO satellite landscape. In addition, and representing the core of this report, are the detailed and specific examples of EO environmental information services across six thematic use cases taken from current ADB operations.

This report is designed to be modular in terms of scope and content. Each chapter can be read as a stand-alone, allowing the reader to focus on specific topics, as necessary. This report is composed of the following:

Chapter 2: Overview of the Satellite Earth Observation Landscape provides an extensive (but not exhaustive) overview of the main categories of players or actors (i.e., the traditional institutional, government, and private sectors, and the rapidly evolving NewSpace sector), the main activities involved in data access and processing, some of the main international or global initiatives that are ongoing, some estimates of the current market for EO data and information services and the projected growth forecasts, the benefits of using EO information for decision-making, and a summary of future trends both in the EO sector and more broadly in the geospatial information sector. The descriptions of companies, organizations, mandates, activities, and programs in this chapter are intentionally based on the descriptions found on the official websites. This is done to preserve the accuracy and consistency of the original sources, and to avoid inventing new descriptions which could introduce misunderstandings.

1 S. Coulson and P. Manunta. 2024. Satellite Earth Observation Assets and Solutions: Supporting Operations in Asia and the Pacific. *ADB Sustainable Development Working Paper Series.* No. 99. Asian Development Bank.

Chapter 3: Use Cases of Satellite Earth Observation in ADB Development Operations is the core of this technical study. This chapter presents a detailed description of six specific and recent cases of EO information being used in ADB development activities. These use cases cover the thematic areas of basic land surface type (cover) mapping and monitoring, disaster risk management (DRM) and disaster risk reduction (DRR), climate resilience and proofing, coastal zone management, forest monitoring and management, and greenhouse gas (GHG) monitoring (CO_2, CH_4, N_2O). Each use case ends with a discussion of how this type of thematic EO information can be further used in development, and how this is linked to the seven ADB 2030 Strategy operational priorities. The chapter concludes with a case example on monitoring changes in the urban environment related to people's access to basic services and adequate, safe, and affordable housing.

Chapter 4: Conclusions briefly summarizes the main conclusions arising from the materials presented in this technical study in terms of the overall EO landscape and the impact of satellite EO environmental information services within the context of ADB development operations.

2 OVERVIEW OF THE SATELLITE EARTH OBSERVATION LANDSCAPE

This chapter contains an extensive overview of the current EO landscape. This broad overview covers the main categories of players or actors (i.e., the traditional institutional, government, and private sectors, and the rapidly evolving NewSpace sector), the main activities involved in data access and processing, some of the main international or global initiatives that are ongoing, estimates of the current market for EO data and information services and the projected growth forecasts, the benefits of using EO information for decision-making, and a summary of future trends both in the EO sector and more broadly in the geospatial information sector.

Although wide in scope, this chapter aims to cover the main issues and developments only. Given this wide scope, it is not possible to be comprehensive or exhaustive, and with the rapid changes taking place in the EO domain, it is inevitable not to include some topics.

It is assumed that the reader has a preliminary understanding of the basic concepts used in remote sensing and satellite EO. If this is not the case, read Chapter 2 of ADB Working Paper No. 99, *Satellite Earth Observation Assets and Solutions* (footnote 1).

▲ Image of Landsat-7 satellite during integration in 1998 (image from National Aeronautics and Space Administration).

EO satellites are traditionally large (i.e., a few thousand kilograms) and expensive (i.e., a few hundred million euros), and have complex engineering systems that take a long time to design, develop, and implement (5–10 years).

These factors dictate that the usual clients for EO satellites are national governments or space agencies, and international agencies such as the European Space Agency (ESA).

Civilian EO began in the early 1970s with the Landsat program of the National Aeronautics and Space Administration (NASA) and the launch of the first Landsat-1 satellite on 23 July 1972.

Since then, the sector has evolved rapidly in terms of technology, capability, and number of countries building and operating EO satellite missions. Europe entered this field in 1986 through the launch of the SPOT-1 satellite via the French space agency (CNES) and the European Remote Sensing Satellite-1 (ERS-1) satellite in 1991 through ESA. Although costs and timeframes are reducing, these satellite missions are still essentially of the same nature (large, expensive, and complex).

Further details on the status of EO in the government sector are provided in Section 2.1.

However, in the last decade, a range of exciting and radical developments have taken place, known collectively as NewSpace. These developments have led to the emergence of small satellite (SmallSat), microsatellite, and nanosatellite disruptive technologies wherein EO satellites (in contrast to the traditional methods) are small (i.e., 1–100 kg) and affordable (i.e., less than 1 million euros, including launch and operations), and have standard engineering systems that are much faster to design and build (i.e., a few months).

This emerging model of NewSpace is characterized by constellations of many micro and nano satellites, built mainly with commercial components and through simplified production and test processes that rely on low complexity. This has led to cheaper manufacturing costs and faster timeframes to put constellations into operation, attracting a wide range of private sector entrepreneurs focused on commercial objectives, and venture capital investments. "As a service" (AaS) business models are now being adopted across the EO value chain, including data analytics, payloads, and ground segments. Therefore, satellite data is usually not free and is available only at commercial rates, but data costs are rapidly falling.

Further details on the status of EO in the NewSpace sector is discussed in Section 2.3.

Figure 1: Evolution in Time of Commercial Satellites in Operation

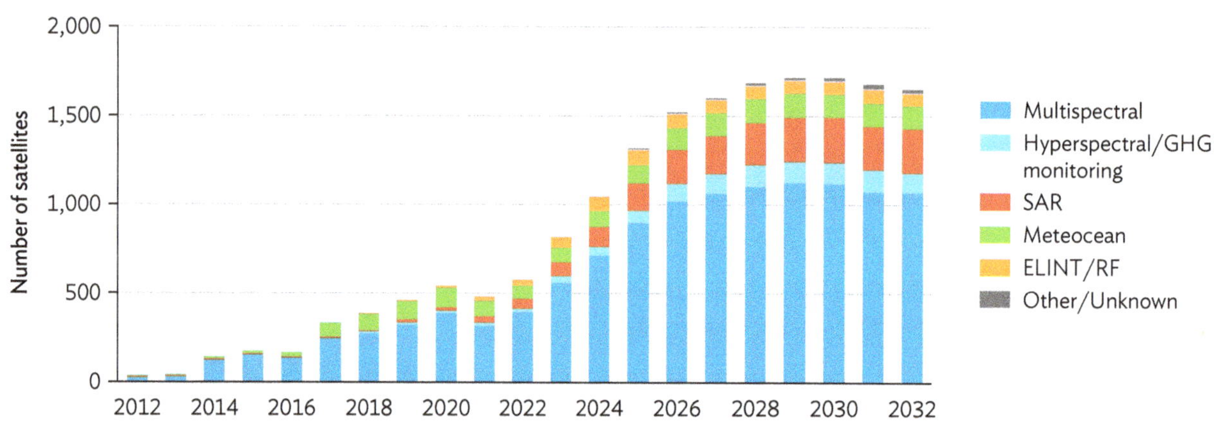

ELINT/RF = electronic intelligence/radio frequency, GHG = greenhouse gas, SAR = synthetic aperture radar.
Source: Euroconsult. 2024. *Earth Observation Data and Services Market.* 17th Edition. November. https://digital-platform.euroconsult-ec.com/product/earth-observation-data-services-market/

As of November 2024, EO data was being sold commercially from more than 800 satellites.[2]

This is forecasted to grow to almost 1,500 satellites in 2033, with this number being composed almost entirely by constellations of optical and synthetic aperture radar (SAR) satellites (with most of these constellations consisting of more than 30 satellites).

[2] Euroconsult. 2024. *Earth Observation Data and Services Market.* 17th Edition. November. https://digital-platform.euroconsult-ec.com/product/earth-observation-data-services-market/.

These numbers include satellites from private enterprises and governments whose data is made available on a commercial basis and exclude noncommercial satellites (e.g., science and public good programs).

Based on current planning by EuroConsult, by 2026–2028, constellations are expected to become more of a replenishment cycle maintaining the required number of satellites in orbit. This forecast is based on current satellite launch schedules and individual operational lifespans. However, the figures could rise further, with new companies emerging later in the decade and with countries commercializing data from their EO programs, particularly the People's Republic of China (PRC).

A more complete listing and description of currently active and historical satellite missions (1,059 entries) is available on the eoPortal managed by ESA.[3]

2.1 Institutional and Government Sector

2.1.1 Intergovernmental Initiatives

Some key intergovernmental initiatives are currently driving and shaping the EO landscape across the planet, but the most substantial programs and activities are being implemented in Europe—the Copernicus program of the EU and the EO programs of the European Space Agency.

These activities are synthesized in this section based on the official sources of information. Other main EO organizations and activities currently taking place globally are then broadly described.

European Union

The most comprehensive, complete, and capable EO satellite system in the world today is Copernicus. Copernicus is the EO flagship program of the EU, which monitors the planet globally, consistently, and frequently. It provides free EO data and thematic informational services based on satellite EO data and in situ (i.e., non-space) data.[4]

Copernicus
Europe's eyes on Earth

The Copernicus program is funded, coordinated, and managed by the European Commission in cooperation with their partners. These partners include the ESA, the European Organization for the Exploitation of Meteorological Satellites (EUMETSAT), and the European Environment Agency. Through this cooperation and investment in space infrastructure, industrial capability, and Earth science, Europe has become one of the leading actors in the world in the field of satellite EO.

The Copernicus system consists of a set of specifically developed satellites (i.e., the Sentinel families), operating together with "contributing missions" (i.e., existing commercial and public satellites). The Sentinel satellites are custom designed to produce Copernicus thematic informational services, which are responding to and are driven by user requirements.[5]

3 eoPortal (accessed December 2024).

4 Copernicus, Europe's eyes on Earth (Copernicus).

5 SentiWiki.

▲ Copernicus Sentinels fleet (image from the European Space Agency).

Since the launch of Sentinel-1A in 2014, the EU has started a long-term program to develop, build, and launch a constellation consisting of almost 20 additional Sentinel satellites in orbit before 2030. In 2023, there were eight Sentinel satellites operating in orbit, from five different families (i.e., Sentinel 1, 2, 3, 5, and 6).

The Sentinels provide an unprecedented operational capacity to observe and monitor the Earth's environment. For instance, when satellites Sentinel 2-A and 2-B operate together, the full global land surface of Earth is imaged at 13 spectral bands and 10-meter resolution every 5 days (depending on the presence of cloud cover).

Six Sentinel expansion missions, to be deployed beyond 2030, are being designed to fill current gaps in user information requirements, and to extend the technical measuring capabilities of the Copernicus system to better address EU policies.[6]

These expansion missions will enable a whole range of new environmental issues to be specifically monitored. This includes monitoring of sea surface temperatures, sea surface salinity, sea ice concentrations, sea ice thickness and overlying snow depth, and changes in the height of ice sheets and glaciers; monitoring of atmospheric carbon dioxide produced by human activity and land surface temperatures; improvement in forest management; monitoring of land subsidence and soil moisture; identification of crop types for precision farming and food security; and sustainable agriculture and biodiversity management.

6 The European Space Agency (ESA). Copernicus Sentinel Expansion Missions.

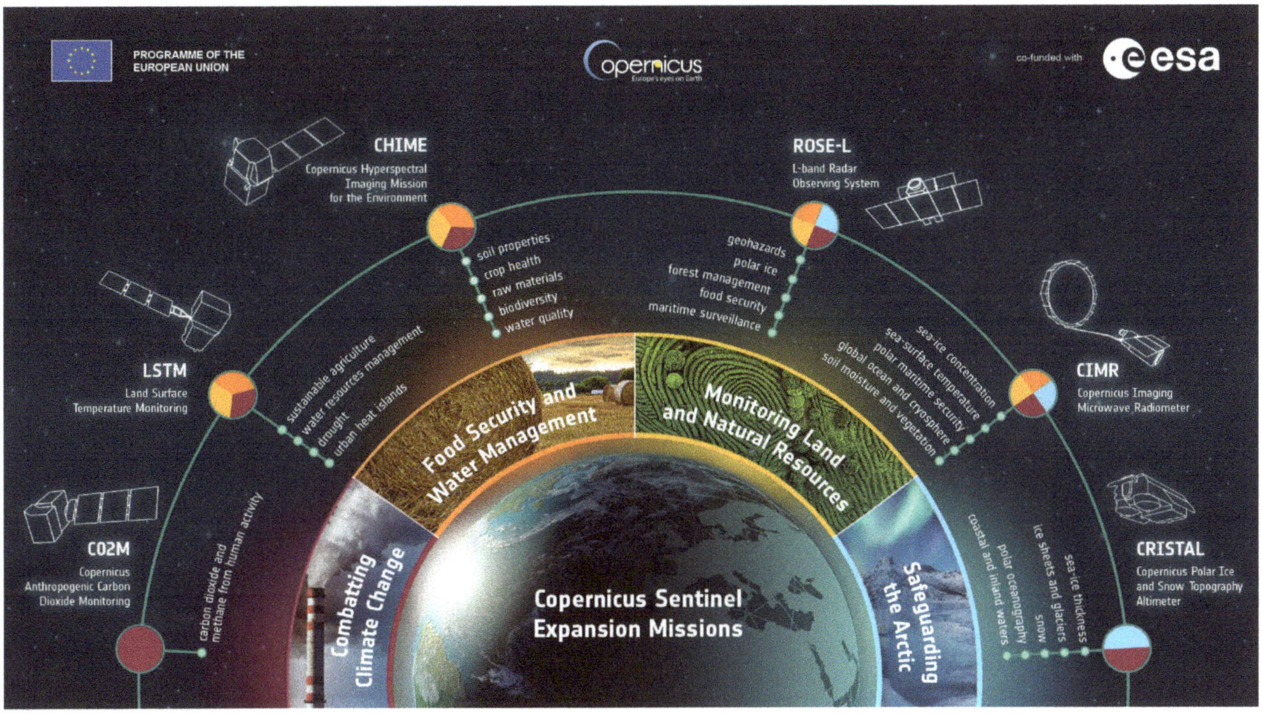

▲ Copernicus Sentinel expansion missions (image from the European Space Agency).

In 2023, Copernicus Sentinel satellites were the largest producers of EO data in the world, providing more than 25 terabytes of global data per day. This huge volume of data is now accessible through the newly developed Copernicus Data Ecosystem, which is the latest development in providing easier access to EO data.[7]

The Copernicus Data Space Ecosystem supports users in discovering, accessing, viewing, downloading, and analyzing EO data. This enables users to build numerous applications that provide accurate, timely, and objective environmental information, thereby creating a more sustainable future.

Box 1

Copernicus Data Space Ecosystem

- Largest Earth Observation data offering in the world, with discovery and download capabilities
- A set of data processing tools to extract objective information and conduct public, private, or commercial activities
- A thriving ecosystem to offer data, services, and applications from public, commercial, and scientific service providers
- A service to benefit institutional users, research and commercial sectors, and every citizen of our planet

Source: European Union.

[7] Copernicus Data Space.

Figure 2: Six Copernicus Services

Source: European Union.

The Copernicus services routinely process these huge daily volumes of basic satellite EO and in situ data. This results in environmental informational products that are more advanced in value than the basic EO satellite data and are more accessible with lower data volumes.[8] These processing activities are implemented within six thematic lines, each producing thematically specific Copernicus informational services (atmosphere, marine, land, climate change, security, and emergency).

Historical datasets offer the ability to research and compare information across years and decades, with data searchable by space and time, which is crucial for monitoring and tracking the changes. Features and anomalies on the Earth's surface can be identified via maps, and extracting statistical information is possible both temporally and spatially. In addition, new patterns can be identified in this huge volume of data and used to create better forecasts regarding the processes impacting the ocean and the atmosphere.

The main users of Copernicus services are public authorities and organizations. They may be involved in operational activities that require accurate and up-to-date environmental information to make more informed decisions, such as in managing an emergency during a disaster event or a humanitarian crisis. They may also use Copernicus information services for better planning and management of their future activities, policies, and strategies.

The data collected through sentinels and contributing missions allows further development of value-added services, which are focused on more-specific public goods or commercial needs. This additional activity generates new business opportunities and more growth in the digital economy.

Most of the Copernicus information services, as well as the EO data from which they are derived, are accessible on a free, full, and open basis by anyone, without restriction. This data and information are used by public authorities, private companies, and international organizations to improve the quality of daily life for citizens both in Europe and globally, to monitor and mitigate climate variability, and to manage and preserve our fragile environment.

[8] Copernicus. Copernicus Services.

European Space Agency

The ESA is one of the global leaders in EO and is at the forefront of developing cutting-edge EO technologies and capabilities to better understand many of the scientific aspects of the Earth, improve citizen's daily lives, support effective environmental policymaking for a more sustainable future, and grow EO-based businesses and the digital economy.

The ESA has more than 30 years of experience in the EO domain in developing countries, taking the lead in Earth science and EO-based applications and informational products. It started with the launch of the pioneering European Remote Sensing Satellite-1 (ERS-1) in July 1991.[9] ERS-1 was the first environmental monitoring satellite developed by ESA, serving as a pathfinder for many of the EO applications that exist today. This mission detected land and ocean surface changes and provided key observations on the Earth's oceans, polar ice, vegetation, geology, meteorology, and ecology.

Today, ESA's Earth Observation Directorate implements a substantial range of development programs and activities encompassing all aspects of EO.[10] These include the following:

- The **Earth Explorer** missions are designed to improve our understanding of key scientific aspects of the Earth. They use advanced space technologies that are often developed and deployed in space for the first time during the mission implementation. Each Earth Explorer focuses on a specific scientific issue regarding the interactions between the atmosphere, biosphere, hydrosphere, cryosphere, and the Earth's interior.
- The **Sentinel** missions are developed specifically for the EU's Copernicus program, the largest EO environmental monitoring program in the world. By providing a set of key operational and informational services on key aspects of the Earth's environment, consistently on a regular basis, the Copernicus program is truly a wholesale change in the way we manage the environment, understand and tackle the effects of climate variability, and safeguard everyday lives.
- **Meteorological** satellites are being operated through a cooperative partnership between the ESA and EUMETSAT. Operating in both geostationary orbits (e.g., the MeteoSat-9, 10, and 11 satellites covering Africa, Europe, and India) and polar orbits (e.g., the two Metop satellites covering the globe), these satellites provide essential information for weather forecasts, issue warnings, and support meteorological decision-making. Information from these satellites is also critical in understanding climate variability.

The ESA also carries out many EO application activities, both in the basic Earth science domain, and through initiatives for applying EO information to real-world problems, such as the Climate Change Initiative, the Earth Observation Science for Society activities, and the International Charter "Space and Major Disasters."

ESA's Φ-lab innovates together with industrial and academic partners, and runs the InCubed program, a co-funding initiative that helps European startups, companies, and entrepreneurs in getting innovative EO products and services to market.

9 eoPortal. 2024. *ERS-1 (European Remote-Sensing Satellite-1)*.

10 ESA. ESA for Earth.

▲ Images sourced from European Space Agency.

Finally, Earth Online contains extensive information on ESA activities in the field of EO.[11] The website presents a wealth of technical information and specifications about ESA's satellite missions, their payload instruments, and the EO data that they acquire.

In addition to these two substantial EO initiatives currently transforming the EO landscape, the following organizations, activities, and capabilities are broadly summarized. Further details can be found through the sources referenced in the footnotes.

European Union Satellite Centre

The European Union Satellite Centre (SatCen) was founded in 1992 as a Western European body and later integrated into the EU as an agency on 1 January 2002.[12]

The SatCen supports the EU in the field of Common Foreign and Security Policy, in particular Common Security and Defense Policy, including crisis management missions and operations.

European Union European Space Programmes Agency

The European Union European Space Programmes Agency (EUSPA) is responsible for managing operations of the European Geostationary Navigation Overlay Service and Galileo satellite navigation programs, ensuring the continuous and reliable provision of their positional and timing services.[13]

The EUSPA also supports the development of integrated applications based on Galileo, the European Geostationary Navigation Overlay Service, and Copernicus data through EU funding mechanisms (e.g., Fundamental Elements and Horizon Europe). The goal is to promote user consumption of data and services to maximize their socioeconomic benefits.[14]

11 Earth Online.

12 European Union Satellite Center.

13 European Union Agency for the Space Programme (EUSPA).

14 EUSPA. Industry Sectors.

European Organization for the Exploitation of Meteorological Satellites

The European Organization for the Exploitation of Meteorological Satellites (EUMETSAT) is the European agency for the provision of operational satellite imagery, data, and services for monitoring weather, climate, and the environment from space.[15]

Numerical weather forecasting requires very frequent and high-quality observational data about the state of the atmosphere in real time. To provide this type of EO data, EUMETSAT operates a number of the geostationary satellites with a permanent presence and continuous coverage of satellite imagery for specific areas of the globe (e.g., Meteosat-10 and Meteosat-11 over Europe and Africa, and Meteosat-9 over the Indian Ocean). It also operates two MetOp polar-orbiting satellites providing global coverage with frequent revisit capabilities.

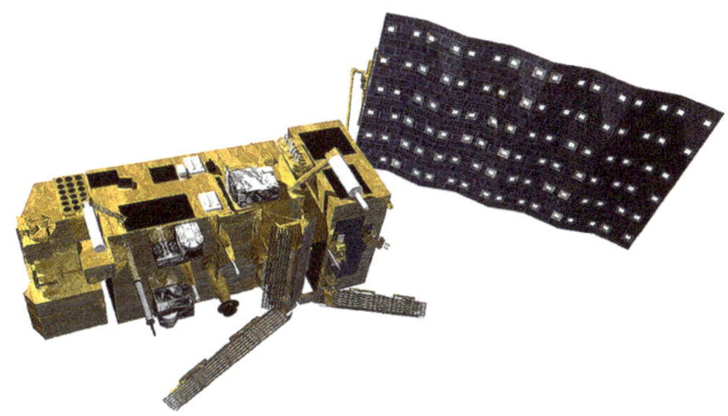

▲ MetOp B and C series satellites (image from EUMETSAT).

They provide vital information in issuing warnings and supporting decision-making in the event of weather, climate, and water-related hazard extreme events, which accounted for more than 2 million human deaths and losses of €3 trillion between 1970 and 2019.

The EU has designated EUMETSAT with processing the data from four of the Copernicus Sentinel missions dedicated to the monitoring of the atmosphere, ocean, and climate. EUMETSAT carries out these tasks in cooperation with the ESA and processes data from Sentinel-3 and Sentinel-6 marine missions.

EUMETSAT has established cooperation with EO satellite operators in the People's Republic of China, Europe, India, Japan, the Republic of Korea, and the United States. The cooperation with the Russian Federation was suspended in March 2022.

Asia-Pacific Regional Space Agency Forum

The Asia-Pacific Regional Space Agency Forum (APRSAF) was established in 1993 to enhance space activities in the region.[16] Space agencies, government departments, international organizations, private sector companies, universities, and research institutes from more than 40 countries actively take part in APRSAF.

APRSAF currently organizes four working groups—Space Applications, Space Technology, Space Environment Utilization, and Space Education—to share information about space activities, policies, future plans, and ideas for technological development in each country and region in these respective areas.[17]

15 EUMETSAT. Who We Are.

16 Asia-Pacific Regional Space Agency Forum (APRSAF).

17 APRSAF. APRSAF Brochure.

Space Applications Working Group (SAWG)

Space Environment Utilization Working Group (SEUWG)

Space Technology Working Group (STWG)

Space Education Working Group (SEWG)

APRSAF also supports the establishment of international projects as solutions to common issues, such as disaster management (Sentinel Asia) and environmental protection (SAFE), so that participating parties will benefit from mutual cooperation.

APRSAF holds an annual meeting in the autumn, which has become one of the "must-attend" space forums in Asia and the Pacific. The last scheduled meeting is the 30th session in the series and it was set on 26–29 November 2024, in Perth, Australia, with a focus on the theme, *"Collaborating to Build a Sustainable and Responsible Regional Space Sector."*[18]

United Nations Satellite Centre

The United Nations Satellite Centre (UNOSAT) is part of the UN Institute for Training and Research (UNITAR). UNOSAT provides geo-information, from satellite imagery analyses to UN programs and specialized agencies. The center also supports UN member states upon request with customized analyses of satellite imagery over their local territories, and provides training and capacity development to build national expertise in the use of geospatial information technologies.

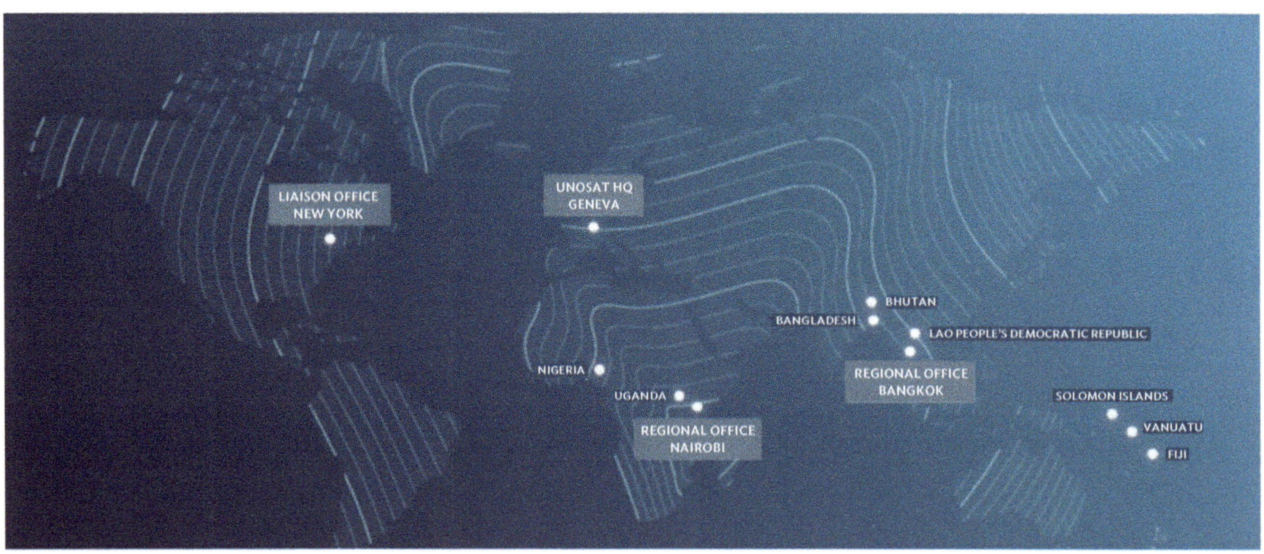

▲ UNOSAT offices worldwide (image from UNOSAT).

[18] APRSAF. Annual Meetings.

UNOSAT provides a range of information services and products focusing on humanitarian emergencies related to natural hazard events, environmental emergencies and conflict situations, human rights violations, and threats to peace and security.[19]

Map 1: Example of Storm Surge Damage Mapping in the Marshall Islands

29 Jan 2024

Marshall Islands (the)

Preliminary Satellite-Derived Damage Assessment over Ailinglaplap Atoll and Arno Atoll, Republic of the Marshall Islands as of 29 Jan. 2024

Storm Surge

Source: Image from UNOSAT.

An example of a national capacity-building project is Strengthening Capacities in the Use of Geospatial Information for Improved Resilience in Asia-Pacific and Africa.[20] Designed to run for 3 years, this project aims to further enhance technical capacities related to disaster risk reduction (DRR), climate resilience, environmental preservation, and food security. It is financed through the support of the Norwegian Agency for Development Cooperation (NORAD) and other funding sources, and implemented by UNOSAT regional liaison offices in Asia and East Africa.

▲ Example of an in-field training (photo by UNOSAT).

United Nations Office for Outer Space Affairs

The United Nations Office for Outer Space Affairs (UNOOSA) is mandated to promote international cooperation for the peaceful use and exploration of space, and the use of space science and technology for sustainable economic and social development.

UNITED NATIONS
Office for Outer Space Affairs

[19] UNOSAT. UNOSAT Services; UNOSAT. Products.

[20] United Nations Institute for Training and Research. Strengthening Capacities in the Use of Geospatial Information for Improved Resilience in Asia-Pacific and Africa.

The UNOOSA supports UN member states in establishing legal and regulatory frameworks to govern space activities, and build the technical capacity of developing countries in using space science and applications technologies for the purposes of development by supporting them with integrating space capabilities into their national development programs. In addition, on 31 January 2024, UNOOSA and Exolaunch GmbH signed an agreement to provide free launch opportunities for Cube Satellites (CubeSats), thereby offering developing countries easier access to space.[21] The development of CubeSats has enabled more affordable and achievable access for these countries when participating in space program activities.

▲ Image from UNOOSA.

The UNOOSA has implemented the Program on Space Applications since its creation in 1971, which has expanded the knowledge and use of space applications in many countries around the world, thereby reducing the digital technology gap that exists between industrialized and developing countries.[22]

The UNOOSA acts as the secretariat of the UN Committee on the Peaceful Uses of Outer Space (COPUOS), with 102 participating member states that meet annually in Vienna to monitor and discuss the developments related to exploration and use of outer space.

2.1.2 Government Initiatives

Developed economies have made substantial investments in satellite EO over the last 50 years. These capabilities contribute to global efforts in environmental monitoring, climate variability assessments, disaster management, and sustainable resource utilization. Collaboration between governments and international organizations has increased data accessibility and sharing, further enhancing the benefits of EO on a global scale.

A few examples of major EO capabilities include the following:

United States

The US operates an extensive portfolio of EO programs globally, led by the National Aeronautics and Space Administration (NASA) and the National Oceanic and Atmospheric Administration (NOAA).

NASA's Earth Observing System (EOS) is a coherent system of polar-orbiting satellites for the long-term global observation of the land surface, biosphere, solid Earth, atmosphere, and oceans.[23] The Earth Science Division operates a variety of satellites, such as Landsat-9, Jason-3, GRACE-FO, Terra, and Aqua, which monitor the Earth's climate, weather, and environmental changes.[24]

In addition, NOAA provides essential weather data through satellites like GOES.

---

[21] United Nations, Information Service Vienna. 2024. *UN Office for Outer Space Affairs and Exolaunch sign agreement to launch CubeSats into space.*

[22] United Nations, Office for Outer Space Affairs. United Nations Programme on Space Applications.

[23] NASA's Earth Observing System, Project Science Office.

[24] NASA Earth Science.

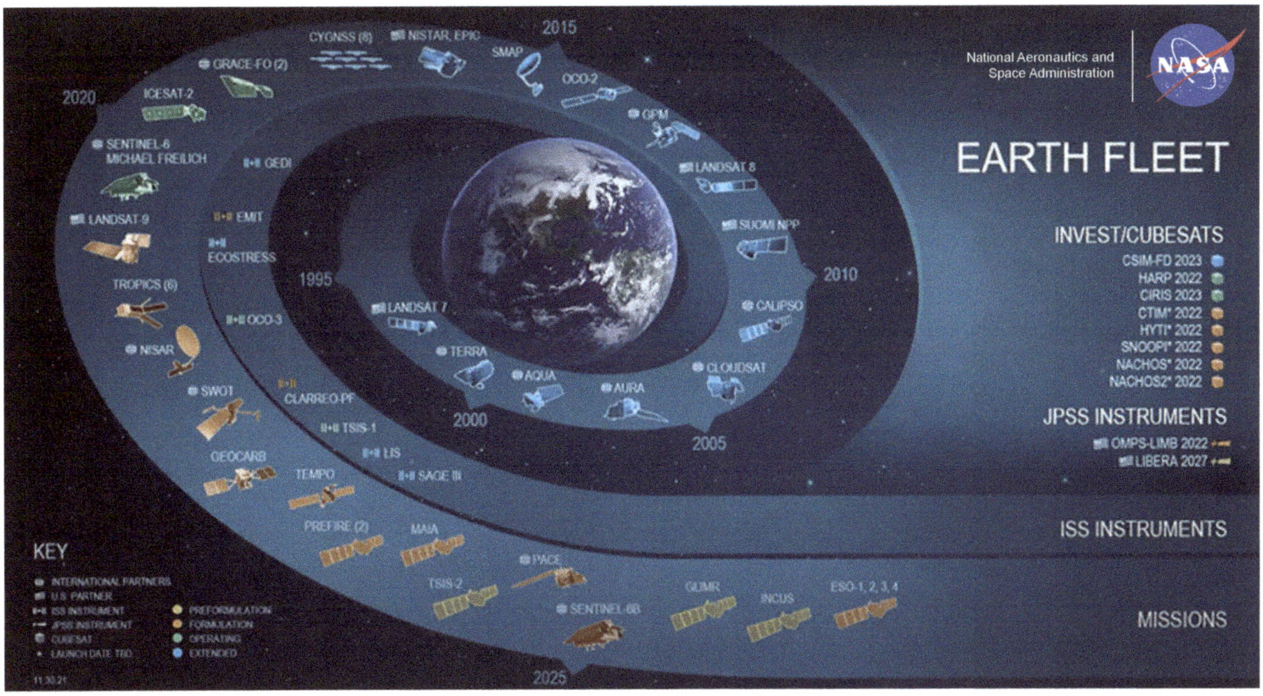

▲ The NASA Earth Science Division Satellite Missions (image from the National Aeronautics and Space Administration).

People's Republic of China

The People's Republic of China (PRC) has made significant developments in the last 20 years in EO with the China National Space Administration.[25] This includes the Gaofen and GaoJing series of optical satellites for EO, the Fengyun series for meteorology, the Haiyang series for oceanography, the Huanjing series for environmental and disaster monitoring, and the China Seismo-Electromagnetism Satellite mission. The pace at which new satellites are being developed and deployed is now greater than that of both the US and Europe.

▲ Gaofen 12 (04) launch from the Jiuquan Satellite Launch Centre in northwest People's Republic of China (image from the China National Space Administration).

Gaofen 12 (04) was launched on 21 August 2023, joining three other Gaofen satellites operating in orbit.[26] Gaofen means "high resolution" in Chinese, and the satellites are operated by the China High-resolution Earth Observation System. These satellites will be used in a variety of applications including land classification, urban planning, road network designs, crop yield estimations, and disaster management.

25 China National Space Administration.

26 Space.com. 2023. China launches Gaofen Earth-observation satellite.

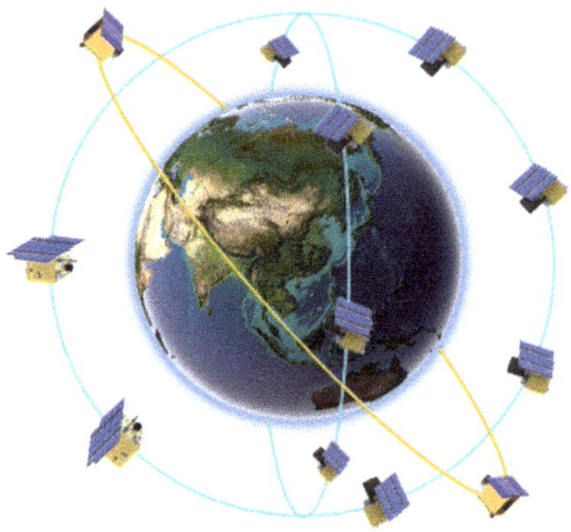

The Zhuhai-1 constellation: 12 satellites currently in orbit, 34 planned in total (image from Orbita).

In addition, the PRC's private sector EO initiatives are now starting with the Jilin satellite constellation (1-meter resolution imagery) and the Zhuhai-1 constellation operated by Orbita which was founded in March 2000 (14 video satellites, 10 hyperspectral, 2 radar, and 8 infrared satellite missions).[27]

Japan

The Japan Aerospace Exploration Agency (JAXA) manages the national EO program,[28] with current satellites in operation that include the Advanced Land Observing Satellite (ALOS-4) launched in July 2024. EarthCare was developed jointly with the ESA, launching the Global Change Observation Mission – Climate (GCOM-C), and the Greenhouse Gases Observing Satellite (GOSAT-2) on 29 May 2024. These satellites monitor greenhouse gases (carbon dioxide and methane), water circulation mechanisms and climate variability, land use, cartography, disaster management, land and ocean surface, and atmospheric measurements related to carbon cycle and radiation budget (e.g., clouds, aerosols, ocean color, vegetation, and snow and ice).

Additional satellites within these series are now under development to enhance functionality and performance (i.e., the GOSAT-GW planned for launch in 2024).[29]

The JAXA Satellite Applications and Operations Center (SAOC) promotes the applications of EO data, and develops, operates, and maintains the various ground systems required for satellite operations including data acquisition, storage, provision, and processing.[30]

In addition, the Japanese Meteorological Agency has been operating geostationary meteorological satellites since 1977. The most recent one in this series is the Himawari-9, which was launched in 2022 and will remain operational throughout 2029.[31]

---

27 Zhuhai Orbita Aerospace Science & Technology Co. Ltd.

28 Japan Aerospace Exploration Agency.

29 Space Technology Directorate I. The Global Observing SATellite for Greenhouse Gases and Water Cycle (GOSAT-GW).

30 Earth-graphy, All about JAXA's Earth Observations.

31 Japan Meteorological Agency. Meteorological Satellites.

▲ The ALOS-4 satellite with phased array L-band imaging radar instrument (PALSAR-3) (image from the Japan Aerospace Exploration Agency).

▲ The Global Observing Satellite for greenhouse gases and water cycle (GOSAT-GW) (image from the Japan Aerospace Exploration Agency).

India

The Indian Space Research Organization (ISRO) has one of the largest constellations of EO satellites in operation, starting with the launch of IRS-1 in March 1988.[32] Since then, India has successfully launched 34 EO satellites featuring the EOS series (supporting agriculture, forestry and plantations, and disaster management), the Cartosat series (supporting urban development and planning, management of rural resources and infrastructure development, and monitoring of coastal land use and land cover), the RESOURCESAT series (supporting natural resources monitoring), and the IRS series (supporting agriculture and land cover). The latest launches are a radar imaging satellite named EOS-4 on 14 February 2022 which supports agriculture, soil moisture and hydrology, and flood mapping, and EOS-6 on 26 November 2022 equipped with an optical imager and scatterometer which scans for ocean colors, wind vector data, and sea surface temperatures.

▲ The launch of EOS-6 together with eight nano satellites from the Satish Dhawan Space Centre in Sriharikota on 26 November 2022 (images from the Indian Space Research Organization).

ISRO provides the satellite services required for weather forecasting and the country's ongoing development and security needs. ISRO also provides data and applications for the public, including information about agriculture, forestry, mining, water management, and rural and urban development, through the Bhuvan and Bhoonidhi Geo-information portals.[33] India displays a thriving and well-established industry offering a wide range of environmental information services based on EO data.

32 Indian Space Research Organization (ISRO), Department of Space. Earth Observation Satellites.

33 ISRO, Department of Space. Space Based Earth Observation Applications.

▲ The EOS-6 first-day images covering the Himalayan region, Kutch region of Gujurat, and the Arabian Sea (image from the Indian Space Research Organization).

Canada

The Canadian Space Agency (CSA) operates the RADARSAT Constellation Mission (RCM). Launched on 12 June 2019, the RADARSAT Constellation consists of three identical radar imaging satellites operating in parallel for maritime surveillance, disaster management, and ecosystem monitoring. The RADARSAT-2 commercially operated satellite was launched on December 2007 for ice, marine, and environmental monitoring, and disaster and resource management. The SCISAT satellite was launched on 13 August 2003 to monitor the ozone layer over Canada and the Arctic.[34]

▲ The RADARSAT Constellation Mission (image from Canadian Space Agency).

France

The SPOT (Satellite pour l'Observation de la Terre) satellite series has been providing high-resolution, wide-area optical imagery of the Earth's surface since 1986. All of the SPOT satellites provide imagery in panchromatic and multispectral bands with a swath width of 60 km.

[34] Government of Canada, Canadian Space Agency. Earth Observation Satellites.

▲ The Pléiades satellite (image from Centre National d'Etudes Spatiales).

▲ Pléiades imagery of Jakarta, Indonesia (image from Centre National d'Etudes Spatiales).

Developed by the Centre National d'Etudes Spatiales (CNES), five satellites were launched between 1986 and 2015, and have pioneered new applications in land surface mapping through land use and land cover, vegetation monitoring, and disaster impact assessments.[35] The most recent satellites in the series (SPOT 6 and SPOT 7) are commercial satellites owned and operated by Airbus Defence and Space. They will maintain the continuity of high-resolution optical data throughout 2024.

The Pléiades constellation, consisting of two satellites, were launched on 17 December 2011 (Pléiades 1A) and 2 December 2012 (Pléiades 1B). The Pléiades system is designed at very high resolution for numerous applications in the fields of mapping and cartography, precision agriculture, forestry, hydrology, and raw material discovery through geological prospecting.[36]

Italy

The Italian Space Agency (ASI) COSMO-SkyMed system is the first EO mission designed at the outset for dual-use purposes (i.e., both civil and military use).[37] As such, COSMO-SkyMed was developed by the ASI in cooperation with the Ministry of Defense.

The COSMO-SkyMed system is a constellation of satellites, with the primary sensor functioning as a high-resolution imaging radar (SAR) which operates in the X-band. As an active system, it can observe the Earth's surface both day and night and in any weather conditions due to the cloud penetration capability of the radar. This radar imagery is used to monitor landslides and floods, coordinate disaster relief efforts, and check crisis zones. The system is operated on a commercial basis, with users able to request COSMO-SkyMed data at a cost through the e-Geos web site.[38]

35 Earth Online. SPOT.

36 Earth Online. Pleiades.

37 Italian Space Agency. COSMO-SkyMed.

38 e-Geos.

▲ COSMO-SkyMed satellite and the Mausoleum of Khoja Ahmed Yasawi, Kazakhstan (images from the Italian Space Agency, e-GEOS).

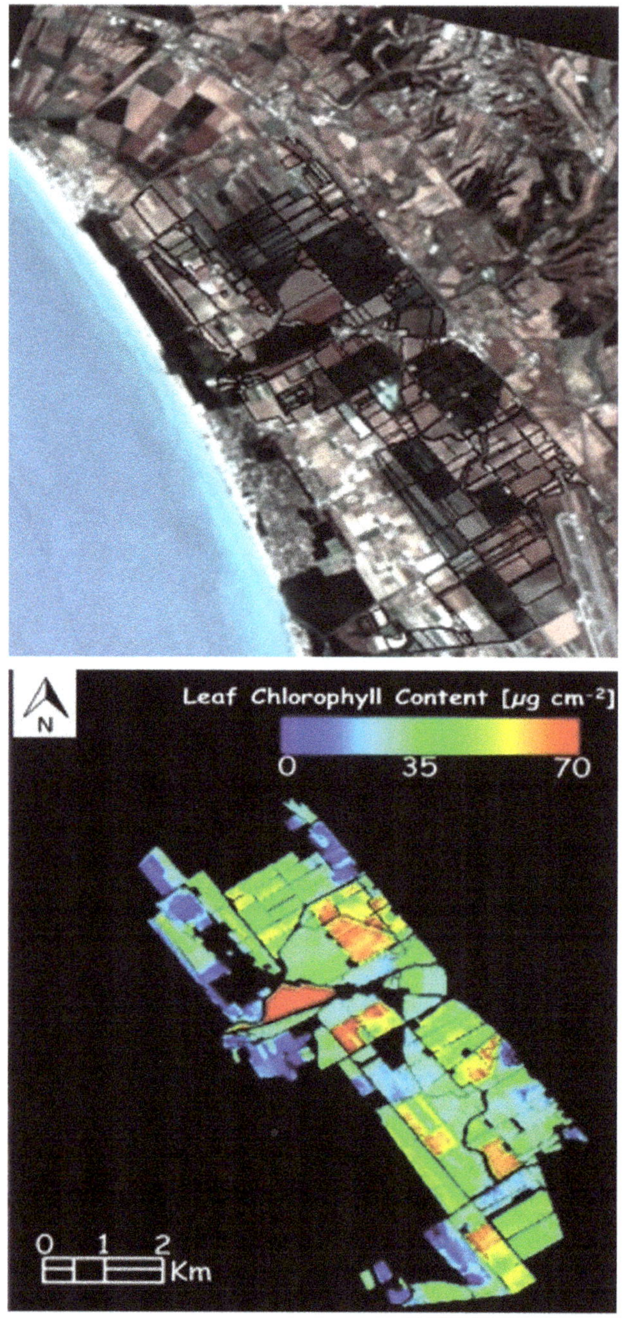

▲ PRISMA hyperspectral imagery for retrieval of vegetation indexes, for the area of Maccarese, an agricultural site located in north of Rome (image by Italian Space Agency, e-GEOS).

The first-generation constellation (four satellites) was deployed between June 2007 and November 2012. The second-generation constellation (CSG-2) started with the deployment of the first enhanced satellite in 2019 and will continue with further launches in 2023–2024.

In addition, ASI operates the PRISMA satellite which carries a hyperspectral (optical) sensor with a medium-resolution camera sensitive to all visible colors (i.e., panchromatic). This combined capability allows the satellite to distinguish both the geometric characteristics of the observed objects and their chemical-physical composition. PRISMA was launched on 22 March 2019 from ESA's spaceport in Kourou, French Guiana.

ASI also distributes data from the Satélite Argentino de Observación Con Microondas (SAOCOM) constellation. These two satellites are equipped with L-band imaging radars and were launched in 2018 and 2020. They are operated by the Argentinian Space Agency (CONAE) and were developed in collaboration with ASI.[39]

Germany

The German Aerospace Center (Deutsches Zentrum für Luft- und Raumfahrt e.V. [DLR]) and EADS Astrium GmbH collaborated through a public–private partnership (PPP) to develop, build, and operate an X-band imaging radar satellite named TerraSAR-X for scientific and commercial use. It was launched on 15 June 2007 from Baikonur, Kazakhstan.[40]

▲ TerraSAR-X and TanDEM-X in formation flight (image from the German Aerospace Center).

This was followed in June 2010 by the TanDEM-X mission, which consists of an additional TerraSAR-X satellite flying in formation in nearly the same orbit slightly behind the first satellite. In this configuration, the two satellites form a system that are fully configurable SAR interferometers in space. By integrating the imagery from both satellites through the technique of SAR interferometry, it was possible to derive a global Digital Elevation Model with a vertical resolution of two meters over 3 years of tandem flight.[41] This "tandem" technique was first demonstrated and pioneered in 1995–1996 by the ESA using the Active Microwave Instrument imaging C-band radars on the ERS-1 and ERS-2 missions.[42]

▲ EnMap Satellite (image from the German Aerospace Center).

In addition, DLR developed and launched the EnMAP hyperspectral (optical) satellite on 1 April 2022. EnMAP is designed to measure biophysical, biochemical, and geochemical variables globally to increase our understanding of biospheric and geospheric processes, and to ensure the sustainability of our natural resources.[43] The satellite is currently in its commissioning phase and will soon begin operations through 2027.

[39] eoPortal. 2025. *SAOCOM (SAR Observation & Communications Satellite).*

[40] The German Space Agency at DLR. TerraSAR-X.

[41] The German Space Agency at DLR. TandDEM-X.

[42] ESA. ERS 1/2 Tandem Mission.

[43] Environmental Mapping and Analysis Program (EnMAP). EnMAP Mission.

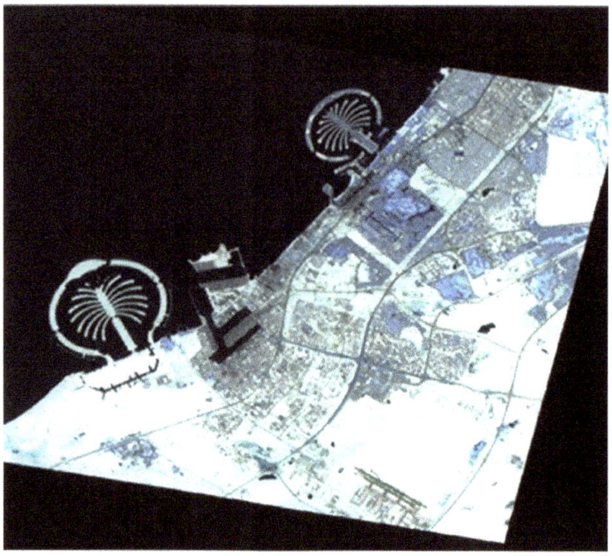

▲ EnMAP imagery of Palm Islands, United Arab Emirates, false color (image from the German Aerospace Center).

▲ DESIS imagery of Haripura Dam, India, November 2022 (image from the German Aerospace Center).

DLR and Germany Teledyne Brown Engineering (TBE) partnered to develop and operate the DLR Earth Sensing Imaging Spectrometer Mission (DESIS) instrument currently installed on the International Space Station.

DESIS measurements significantly improve the ability to measure vegetation health and stress, water quality, pollution, and mineral resources.

Spain

The Agencia Espacial Española is an agency of the Government of Spain, responsible for the Spanish space program. The agency, which became operational in April 2023, consolidates all the space powers of the General State Administration within a single body. They are the National Institute for Aerospace Technology, which was created in 1942, and the Centre for the Development of Industrial Technology, which was established in 1977.

The space programs managed and promoted by Agencia Espacial Española provide key developments in many sectors such as energy, climate, environment, security, defense, health, agriculture and food policies, transportation, tourism, and mobile communication. These activities not only improve regional policies and local planning but also strengthen the economy and increase digital expertise.

PAZ is an X-band synthetic aperture radar (SAR) technology satellite designed to address dual-use applications (i.e., security and defense requirements of the Ministry of Defense and civilian needs). It is capable of taking more than 100 images of up to 25 centimeters (cm) of resolution both day and night, and in all weather conditions.

PAZ provides EO capabilities for multiple purposes: border control, intelligence, military operations, enforcement of international treaties, environmental monitoring, protection of natural resources, land surface motion monitoring, city and infrastructure planning, and monitoring of natural hazard events.

▲ PAZ Satellite and radar imagery of Rome (image from the European Space Agency).

▲ GEOSAT-2 satellite (image from the European Space Agency and Elecnor).

The GEOSAT series (GEOSAT-1 launched in July 2009, and GEOSAT-2 launched in June 2014) are fully owned and operated by the GEOSAT company.

GEOSAT-2 is designed to provide cost-effective and rapid response services delivering fast access to the growing demand for sub-metric resolution imagery. The system provides near-real-time image tasking, data acquisition, processing, and information delivery to the end user.

The High-Resolution Advanced Imaging System (HiRAIS) is a 75 cm multispectral optical instrument with a very-high-resolution camera of five spectral channels (one panchromatic, four multispectral). It has a total imaging capacity of more than 150,000 square kilometers (km^2) per day, with revisit time to the same area on the Earth's surface of 2 days on average globally and even reaching 1 day at mid-latitudes.

In summary, developed economies have made substantial investments in satellite EO, with Copernicus being a key driver, and with data availability and accessibility now growing at an exponential rate. These capabilities greatly contribute to global efforts in environmental monitoring, disaster management, and sustainable resource utilization.

Developing economies in Asia and the Pacific have made significant strides in satellite EO, harnessing space-based technologies for a wide range of applications, including disaster management, agriculture, resource monitoring, and environmental protection. The following provide an overview of satellite EO capabilities in Asia and the Pacific.

▲ GEOSAT-2 image of Busan Harbor, the largest port in the Republic of Korea, acquired in 2017 (image from UrtheCast Corp and Deimos Imaging).

Taipei, China

Taipei,China's space agency, which was officially established on 1 January 2023, has a strong history with EO optical satellites beginning in January 1999 with FORMOSAT-1, operating two instruments: the Ionosphere Plasma and Electrodynamics Instrument and the Ocean Color Imager. This was followed by FORMOSAT-2 in May 2004, operating two instruments: the optical imager Remote Sensing Instrument and the Imager of Sprites and Upper Atmospheric Lightning. The FORMOSAT-3 constellation of six microsatellites came next, operating the GPS Occultation Experiment, which receives the radio wave signals of existing US GPS satellites. Temperature, pressure, and water content in the ionosphere and atmosphere can be derived based on the GPS signal transmission delay caused by electric density.

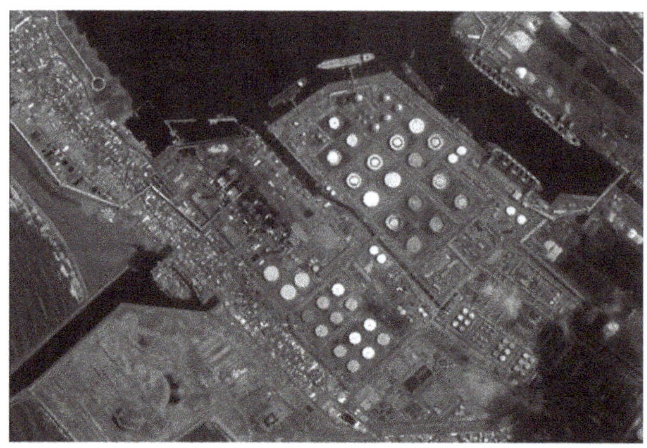

▲ FORMOSAT-2 image of Kao Hsiung, Taipei,China, July 2004 (image from Taipei,China's space organization).

Currently active satellites in operation (all on a commercial basis) in the FORMOSAT series that are used for disaster monitoring, environmental surveillance, and agricultural management are as follows:

- FORMOSAT-5 operates an optical Remote Sensing Instrument, which provides 2-meter (m) resolution panchromatic (PAN, black and white) and 4 m resolution multispectral (MS, color) images.
- The FORMOSAT-7 program is a collaborative space program between Taipei,China and the US (NOAA). FORMOSAT-7 is also called COSMIC-2 (Constellation Observing System for Meteorology, Ionosphere and Climate). It is a constellation of six satellites that are upgraded versions of the

FORMOSAT-3 series. The US is responsible for satellite launches and data acquisition via a global network of ground receiving stations. Taipei,China is responsible for the design and development of the satellites, and mission operations. The FORMOSAT-7 satellites were launched by the SpaceX Falcon heavy rocket on 25 June 2019 at the Kennedy Space Center based in Florida, US.

- The FORMOSAT-8 program is currently under development and consists of six optical remote sensing satellites with a very high resolution of 1 m, which are upgraded versions of the FORMOSAT-5 series. The constellation will provide daily revisit and stereoscopic imaging capabilities to achieve real-time monitoring with global coverage. The constellation will be deployed at the rate of one satellite launch per year over 2024–2029. Research institutes and industry will be involved in the technology development with the aim of reducing satellite costs via the use of commercial off-the-shelf components to save labor and time.

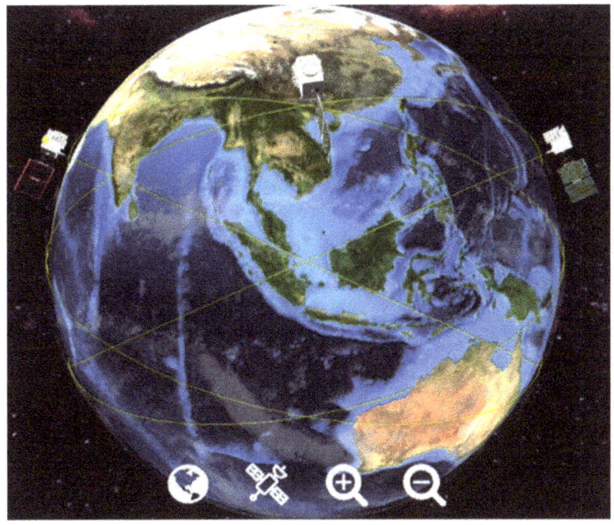

▲ FORMOSAT-7 Satellite Constellation (image from Taipei,China's space agency).

▲ TRITON Weather Satellite (image from Taipei,China's space agency).

On 9 October 2023, Taipei,China's first domestically developed weather satellite, Triton, was launched on-board a Vega rocket by Arianespace from French Guiana. TRITON carries the Global Navigation Satellite System-Reflectometry (GNSS-R, independently developed by Taipei,China's space agency), which analyzes the GNSS signal reflected by the ocean surface to estimate sea wave height and sea surface wind speed. This information is used to better understand air–sea interaction, and improve the prediction of typhoon intensity and direction, which occurs every year in Taipei,China.

Cooperating with local companies to develop an image processing system is one of the development programs established and led by Taipei,China's space agency. The main function of the satellite image processing center is for FORMOSAT satellite image production, including image acquisition and processing. Data Cube has been established by Taipei,China's space agency to manage and operate the data acquired by FORMOSAT series satellites and other open-source remote sensing satellites via cloud computing capabilities.[44]

44 Taipei,China's space agency. Data Cube Application Service.

Philippines

The Republic Act No. 11363 (or the Philippine Space Act), signed into law on 8 August 2019, established the Philippine Space Policy and the Philippine Space Agency (PhilSA).

▲ DIWATA-1 (left) and DIWATA-2 (right) (image from Philippine Space Agency).

In recent years, the Philippines has made significant advances in space science and technology and its applications.

These activities have led to the development, launch, operation, and use of the first domestic microsatellites for scientific EO. These are DIWATA-1 (launched in 2016, mission completed in 2020); DIWATA-2 (launched in 2018 and currently operating); and MAYA-1, the country's first nanosatellite (also launched in 2018, mission completed in 2020).

▲ PhilSA Director General Joel Joseph Marciano Jr. with the MAYA-1 nanosatellite (photo by Philippine Space Agency).

DIWATA and MAYA provide an excellent starting point for the further innovation and evolution of small satellites. These achievements have enabled the Philippines to join the list of economies that are able to build, develop, and operate satellites by themselves.

In addition to these domestic EO satellites, PhilSA also accesses and processes EO data from free sources (e.g., LandSat-8, 9, Sentinels), from satellites in the International Charter Space and Major Disasters (e.g., RadarSat Constellation, Goafen-2, Kanopus-V), and from commercial sources (e.g., Kompsat-5, SPOT 6 and 7, Worldview 2, 3, and 4).

These activities are complemented by a program of ground infrastructure developments including the multi-mission satellite ground receiving stations of the Philippine Earth Data Resource Observation (PEDRO) Center, together with high-performance computing facilities for processing, archiving, and distribution of satellite images.

The Remote Sensing and Data Science Help Desk (DATOS) processes satellite image data and provides environmental informational services for a wide range of applications including agriculture, disaster risk management, and monitoring of marine and maritime domains.

PhilSA is implementing a range of programs and projects, building on existing national technical capabilities, infrastructure, policies, and human capital.[45] These include the Advancing Core Competencies and Expertise in Space Studies Nanosat Project; PhilSA Advanced Degrees for Accelerating Strategic Space R&D and Applications Scholarship Program; Expanding CubeSat Research and Development Efforts in Philippine Universities Initiative; and Space Information, Know-How, and Applications Acceleration through Promotion and Training.

PhilSA is a partner of the EU and ESA in the National Copernicus Capacity Support Action Program for the Philippines, signed in January 2023, to implement a Copernicus data mirror site within PhilSA, to boost the flow of European EO data to the region.[46]

Figure 3: Philippine Space Agency Key Development Areas

Key Development Areas

Source: Philippine Space Agency.

Republic of Korea

Established in 1989, the Korea Aerospace Research Institute (KARI) is the aeronautics and space agency of the Republic of Korea. KARI conducts a wide range of activities in new exploration and technological advancements, development, and dissemination in the field of aerospace science and technology.

For EO, the research and development of satellites in the Republic of Korea began in 1994 with a satellite development project (Arirang). The multi-purpose Arirang-1 optical satellite was developed in 1999, with Arirang-2 developed in 2006. These initial satellites were followed by improvements in the optical series with Arirang-3, Arirang-5, and Arirang-3A (optical + IR) which were developed in 2012, 2013, and 2015, respectively.

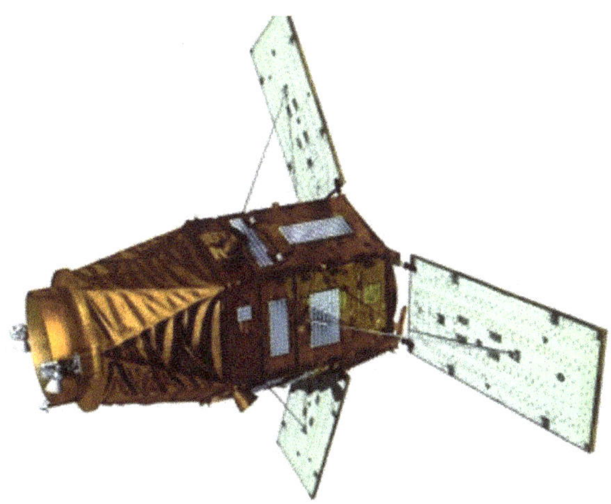

▲ Arirang-3A Satellite (image from Korea Aerospace Research Institute).

Currently, KARI is developing Arirang-6, a high-precision radar satellite, and Arirang-7 and Arirang-7A as high-performance optical satellites equipped with ultra-precise optical and infrared sensors.[47]

45 Philippine Space Agency.

46 Republic of the Philippines, Department of Science and Technology, Advanced Science and Technology Institute. 2023. *First in the Region: Philippines, EU join forces for CopPhil Implementation, strengthens disaster response.*

47 Korea Aerospace Research Institute (KARI). Satellite.

Although the Republic of Korea started relatively late with satellite development in the 1990s, it has progressed rapidly due to extensive investments and a continuous research and development (R&D) program and is now among the world's top six or seven nations in terms of satellite development capabilities. Arirang-3 and Arirang-3A are currently in operation, providing high-resolution optical imagery at 0.7–3 m resolution. Arirang-5 and Arirang-6 are also currently in operation and are all-weather imaging radar satellites with imagery at 0.5–20 m resolution.

In addition to the Arirang series, KARI has also developed the Cheollian series focusing on meteorological and ocean observation services. This series began with the development of Cheollian-1, the nation's first geostationary orbit satellite, followed by Cheollian-2A, which is currently operating and delivering more precise meteorological observations. In addition, Cheollian-2B is delivering marine atmospheric environmental observations from geostationary orbit. Cheollian-2B will track the presence and changes of air pollutants such as fine dust around the Korean Peninsula and is expected to help address social problems stemming from poor air quality.

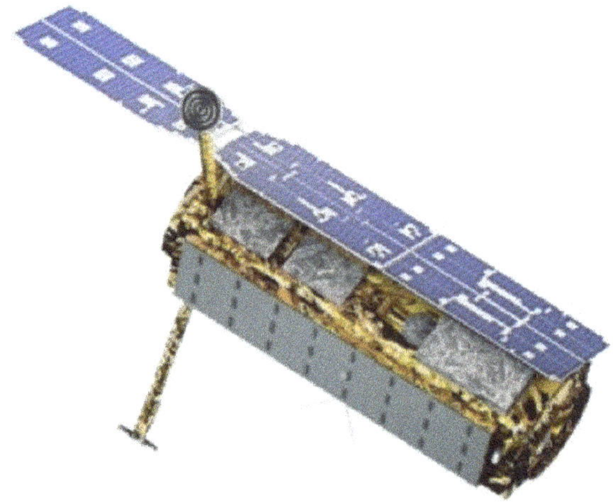

▲ Arirang-6 Satellite (image from Korea Aerospace Research Institute).

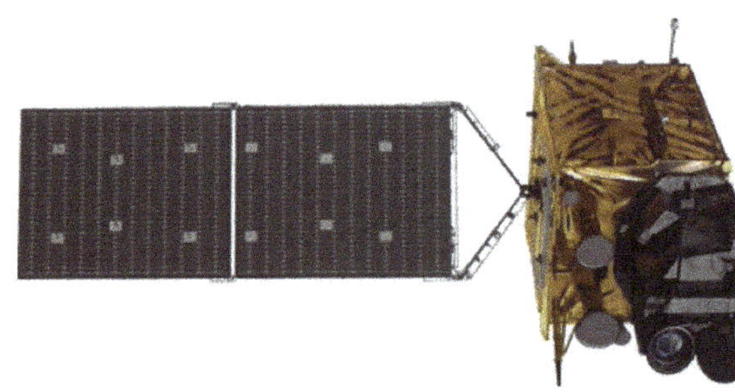

▲ Cheollian-2A Satellite (image from Korea Aerospace Research Institute).

▲ Example imagery from Arirang-3 satellite (image from Korea Aerospace Research Institute).

▲ Example imagery from Cheollian-2A satellite (image from Korea Aerospace Research Institute).

KARI operates the National Satellite Information Utilization Support Center, which supplies satellite imagery to 31 government agencies for public requirements.[48] The center promotes the use of satellite images acquired from the Arirang and Cheollian families in various fields such as land management, disaster monitoring, agricultural and forestry applications, marine water resource management, environmental weather observation, the impact of industrial activities, and national security.

KARI promotes strategic partnerships not only with countries that have developed space programs, but also with those that are beginning to explore their own space-based efforts in the region.

Since 2010, KARI has developed and deployed the KARI International Space Training Course, which is dedicated to building technical capacity in developing countries to strengthen and consolidate these regional space partnerships.[49]

Indonesia

On 1 September 2021, the National Institute of Aeronautics and Space (LAPAN) was transformed into the formation of an integrated agency: the National Research and Innovation Agency (BRIN). In March 2022, the Indonesian Space Agency (INASA) was formed as a special body under BRIN. INASA is the body through which space-based political functions are carried out, including coordination of policies, political lobbying, and collaboration with other space agencies.

Indonesia is one of the largest archipelagoes in the world. It comprises more than 17,500 islands, with 3 million km^2 of the country's territory covered by the sea, and 2 million km^2 of the country's land surface covered mostly by forest and agriculture.

The EO program of LAPAN has been established to use EO and meteorological satellite data operationally, with a focus on supporting disaster management and environmental monitoring. Their aim is to manage and minimize the environmental, social, and economic impacts of natural hazard events, which frequently occur in the region.[50]

It does this by supporting central and local governments in the disaster management sector with up-to-date spatial information to support decision-making, particularly for prevention, preparedness, emergency responses, rehabilitation, and reconstruction.

In support of EO, the Indian Space Research Organization (ISRO) launched on 10 January 2007 the LAPAN-A1/TUBSAT, which was the first Indonesian Surveillance Micro-Satellite (56 kg). It was made by the experts from LAPAN and the Technical University Berlin in Germany and was used for Indonesia's natural resources observation and weather forecasting for the region.

This was followed by LAPAN-A2, an EO microsatellite, which was the first domestic satellite designed and developed by LAPAN and funded nationally by the Government of Indonesia.

48 KARI. Use of Satellite Information.

49 KARI. At a Glance.

50 National Research and Innovation Agency (BRIN) Indonesia. 2024. Satellite Constellation for National Development | APRSAF Talk Show. YouTube video. 23 September.

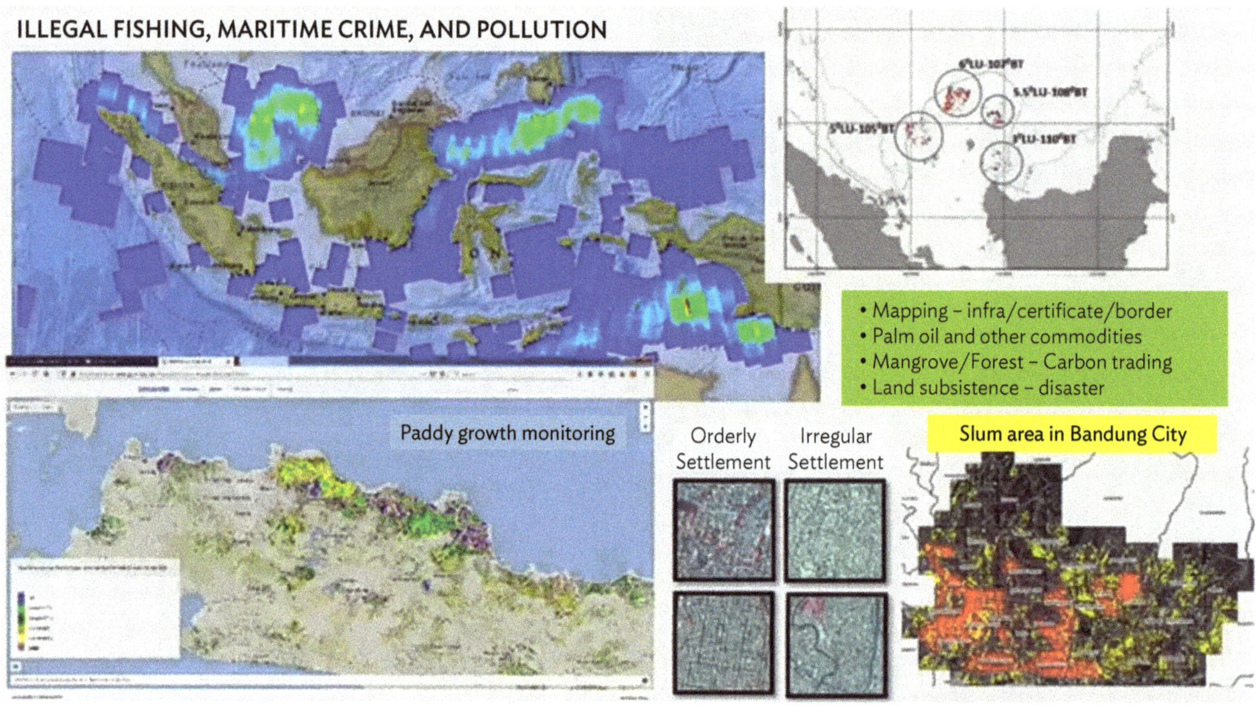

▲ EO applications in Indonesia from the Satellite Constellation for National Development, presented at the Indonesia Research and Innovation Expo InaRI, 20–23 September 2023 (image from Indonesia's National Research and Innovation Agency).

Launched in September 2015, LAPAN-A2 is still operational, and it carries a video camera using commercial-off-the-shelf technology. RGB medium format components are used to support "snapshot" operations. LAPAN-A2 aims to monitor the environment, disasters and natural hazard events, land use, and natural resources.

LAPAN-A2 also carries the automatic identification system (AIS) that uses GPS and VHF digital communication to monitor maritime traffic (position, velocity, and operator of ships). It also carries the Automatic Packet Reporting System, which is an amateur radio voice repeater developed for use by Indonesian Amateur Radio Organization. The Automatic Packet Reporting System provides a satellite-based telecommunication system in support of disaster mitigation and relief efforts during natural hazard events such as earthquakes, tsunamis, volcanic eruptions, and floods.

▲ LAPAN-A1 Satellite (image from Indonesia's National Research and Innovation Agency).

LAPAN-A3/IPB is a microsatellite (115 kg) launched on 22 June 2016 by ISRO in India. LAPAN-A3/IPB carries a medium-resolution multispectral imager (4-band) and a digital camera. It conducts global maritime traffic monitoring, specifically the AIS signals of ships. The mission aims to provide timely, frequent, and accurate optical imagery for monitoring and predicting the condition of food resources and the environment across the Indonesian archipelago. The Bogor Agricultural University (Institute Pertanian Bogor [IPB]) is responsible for algorithm and application development.

Indonesia plans to complete the five microsatellite constellations with the launch of LAPAN-A4 and A5 in 2024, both satellites weighing about 100–300 kg. The primary mission of LAPAN-A4 is to build upon LAPAN-A3 with improved multispectral imagery and AIS measurements. This is complemented by additional missions to support scientific research using a space-based magnetometer and an experimental thermal infrared sensor. LAPAN-A5 will be Indonesia's first microsatellite to carry a synthetic aperture radar (SAR) imaging instrument, which is being built for LAPAN by Chiba University in Japan.

Following the implementation of LAPAN-A4 and A5, Indonesia will launch the Nusantara-5 satellite (SATRIA-2) and the NEO-1 satellite in 2024 on an EO mission to conduct maritime surveillance, and to perform enhanced measurements of the earth's magnetic field.

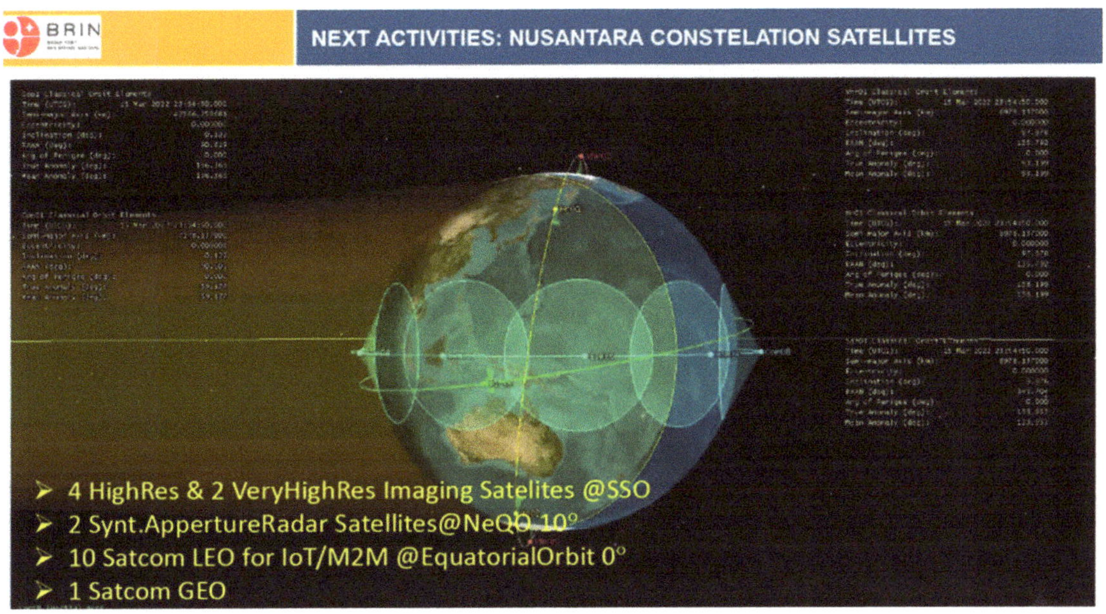

▲ Nusantara concept (image from National Research and Innovation Agency, BRIN).

Indonesia will also deploy the Timau National Observatory in mid-2024 to support space research and technology development.[51]

The Nusantara satellite constellation[52] is to be expanded with the addition of 4 high-resolution and 2 very-high-resolution optical imaging satellites in polar orbit for land mapping; 2 SAR imaging satellites in near-equatorial orbit for frequent maritime surveillance; and 10 low-Earth-orbit communications/Internet-of-Things satellites acting as a data collection platform for disaster sensing (tsunami, earthquakes, volcanoes) and for monitoring air traffic and asset tracking (see the talk show "Satellite Constellation for National Development" presented at the Indonesia Research and Innovation Expo, InaRI, 20–23 September 2023).

51 BRIN. *2024. BRIN Presents Indonesia's Plan to Launch the Nusantara-5 Satellite and Timau National Observatory at the 61st UNCOPUOS Session.*

52 Nusantara Concept presented at Earth Observation National Spatial Data Infrastructure Digital X, ADB event, 19 September 2023. https://events.development.asia/learning-events/digitalxadb-driving-digital-development-across-asia-and-pacific-2023.

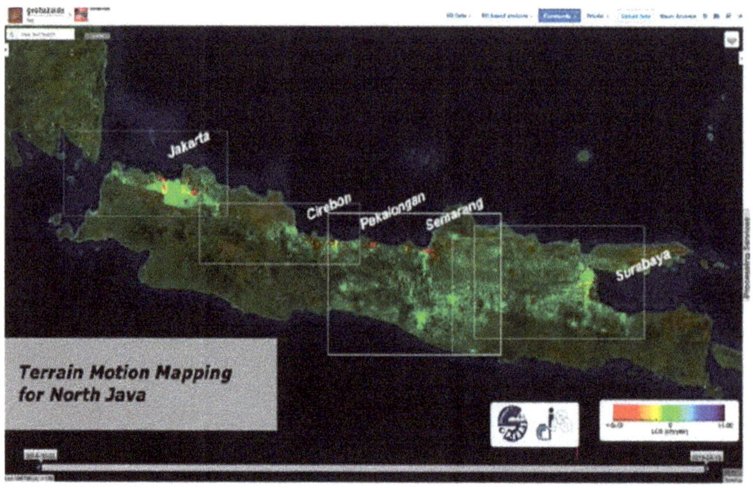

◀ Example of EO Information Processing Platform: Land Surface Motion for North Java (image from Indonesia's National Research and Innovation Agency, Asian Development Bank, and European Space Agency).

For EO applications, BRIN is currently developing a series of cloud-based processing tools in collaboration with ADB. The platform will have a range of dedicated users (ministries, agencies, and local government) who can access the EO information. This collaboration provides support for the "Build Back Better" infrastructure of emergency assistance for rehabilitation projects in Central Sulawesi; for water and food security planning and investments in Indonesia; and for the transfer of EO technology, knowledge, and capacity to Indonesia. The next activities to be addressed over 2023–2027 are for flood management and coastal protection.

In addition to Indonesia's national missions, BRIN acquires a wide variety of optical and SAR data from European and US missions. BRIN increased international cooperation in 2023 by signing a cooperation (memorandum of understanding) with the United Arab Emirates Space Agency, and changes in implementation with the Japan Aerospace Exploration Agency (JAXA).

For many developing countries, bilateral or multilateral collaboration is a favored option to mitigate the technological and financial risks of developing and implementing individual national space programs. Such collaboration allows nations to engage in EO satellite development on a scale that they cannot undertake alone.

Examples of such collaborative missions are as follows:

China–Brazil Earth Resources Satellite.
The China–Brazil Earth Resources Satellite (CBERS) is a technological cooperation program between Brazil and the PRC that develops and operates EO satellites. The program began with the development and deployment of two satellites, CBERS-1 (launched in October 1999) and CBERS-2 (launched in October 2003). A third satellite (CBERS-2B) was added to the program in September 2007. Subsequently, the program expanded with the launch of three additional satellites: CBERS-3 (December 2013),

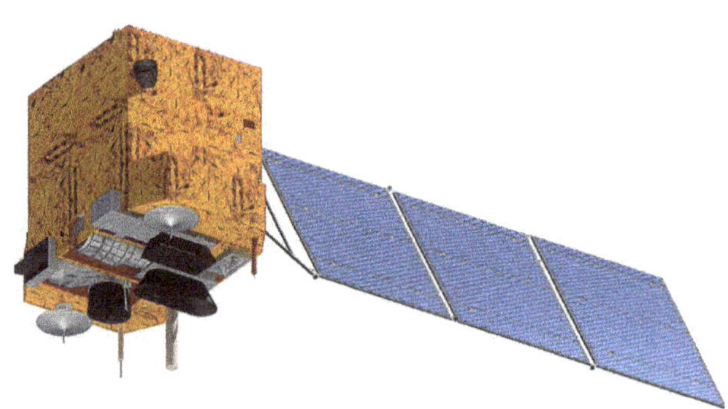

▲ CBERS-4A Satellite, image from Brazil National Institute of Space Research (INPE).

CBERS-4 (December 2014), and CBERS-4A (December 2019). All the satellites carry imaging multispectral radiometers (vis/IR). The main objectives of the mission are to measure land surface topography, Earth surface albedo, vegetation type, and land surface imagery.[53]

NovaSAR-1. Launched in September 2018, this small satellite is a technology demonstrator designed and developed to test the capabilities of a new low-cost S-band SAR instrument with a projected operation period of 7 years.

▲ NovaSAR-1 Satellite (image from Survey Satellite Technology Ltd).

The spacecraft and payload were developed through a collaboration between Survey Satellite Technology Ltd. (SSTL) and Airbus Defense and Space Ltd., with funding from organizations in Australia, India, the Philippines, and the United Kingdom (UK). The satellite is designed to measure vegetation canopy (cover), vegetation type, land cover, sea-ice cover edge and thickness, oil spill cover, and aboveground biomass.[54]

Disaster Monitoring Constellation. The Disaster Monitoring Constellation (DMC) is a unique EO satellite constellation that delivers high-frequency imaging around the globe via a well-established collection of microsatellites built by SSTL.[55] SSTL is a pathfinder of microsatellite technologies with a long history of experience starting from the early 1980s.

▲ DEIMOS-1 Satellite (image from Survey Satellite Technology Ltd).

The DMC constellation began with Alsat-1B in 2002 (Algeria's first EO satellite), BILSAT-1 in 2003 (the first Turkish Scientific EO satellite), UK-DMC-1 in 2003 (UK Space Agency), NigeriaSat-1 in 2003 (Nigeria Space Research and Development Agency), and Beijing-1 in 2005 (China National Space Administration). Each DMC satellite is independently owned and controlled by a DMC consortium member.

The constellation grew in imaging capacity and enhanced performance in 2009 with the addition of the UK-DMC-2 and Deimos-1 satellites (Elecnor Deimos Spanish aerospace company), and added NigeriaSat-2 in 2011 along with the SSTL S1-4 satellite in 2018. The DMC constellation offers rapid coverage of large geographic areas and very frequent revisit worldwide (typically 1–2 days).

[53] eoPortal. 2019. *CBERS-4A (China–Brazil Earth Resources Satellite-4A).*

[54] eoPortal. 2024. *NovaSAR-1.*

[55] Surrey Satellite Technology Ltd (SSTL). The Disaster Monitoring Constellation.

In summary, many economies in Asia and the Pacific have advanced capabilities in the development, production, and operation of microsatellites and CubeSats. They are using EO information on an operational basis for environmental management, disaster risk management, and climate resilience with national ministries and departments. This represents an excellent foundation to consolidate and expand regional collaboration to further grow the use of EO in the region.

2.2 Private Industrial Sector

There are several sources listing the major players in the EO industry.[56]

A useful organizational overview of some EO industries according to their role in the EO value chain can be found in the TerraWatch Space insights, *The State of Commercial Earth Observation, 2022 Edition.*[57]

The EO industry can be classified into three major layers based on its primary function within the EO "operating stack," equally referred to as the EO value chain. These three layers are acquisition, dissemination, and intelligence.

- **Acquisition** is concerned with initially developing the technology of the satellites and their sensors to create remote sensing data and then collecting and organizing it so that it is ready to be accessed and used.
- **Dissemination** is focused on the digital infrastructure and services that make satellite data discoverable, accessible, interoperable, and integrated with other data sources into "analysis ready data" for further processing into environmental information that can be used for a specific application.
- **Intelligence** is where satellite-based environmental information is further transformed into information and insights, and the value is realized by end users, who are operating outside the domain of the EO sector.

Over the last few years, there has been a significant growth of new companies in the acquisition layer. Driving factors in this layer include the following:

- Increased miniaturization of satellites and their subsystems, and use of off-the-shelf components leading to lower costs and easier use of space, with more companies entering the field.
- Innovative business models such as "space as a service" and "satellite as a service," with greater customer focus, thereby reducing the technical barriers to accessing EO data and information.
- A combination of the effects of climate variability, the increasing emphasis on "green and digital transitions" now taking place, and the evolving geopolitical situation increasing demand for security and situational awareness data. These are all leading to greater demand for more data about the planet and the impact of our activities on it.

[56] A. Banerjee. 2025. *Straits Research, Market Research Report*; IMARC Group. 2025. *Satellite-Based Earth Observation Market Report by Solution (Data, Value Added Services), End User (Defense and Intelligence, Infrastructure and Engineering, Agriculture, Energy and Power, and Others), and Region 2025–2033*; Mordor Intelligence. 2025. *Satellite-Based Earth Observation Market Size and Share Analysis: Growth Trends and Forecasts (2025–2030).*

[57] Aravind. 2022. The State of Commercial Earth Observation. *TerraWatch Space*. 28 August.

Figure 4: Examples of Earth Observation Companies Organized Based on Their Primary Function

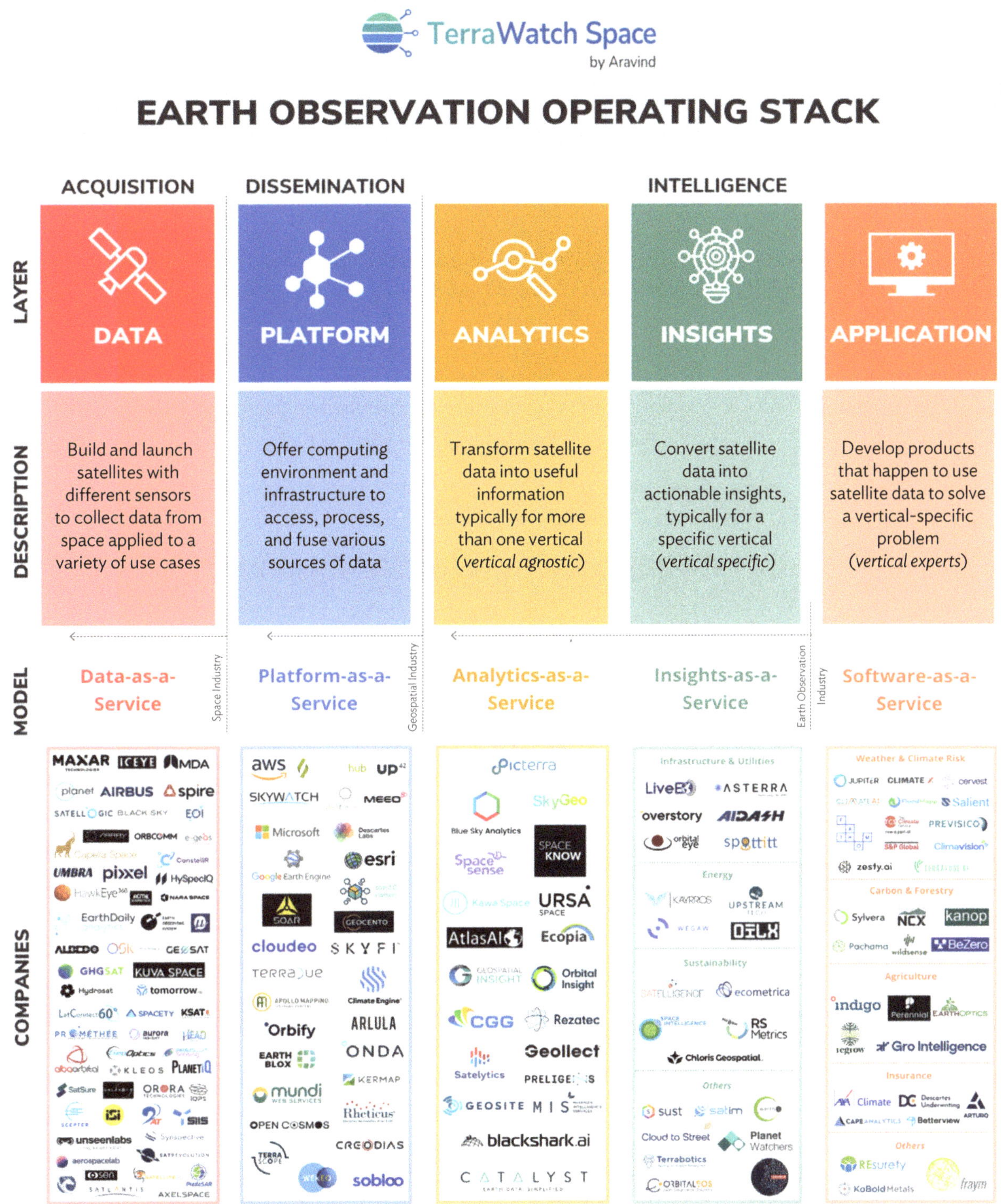

Source: Aravind. 2022. The State of Commercizal Earth Observation: 2022 Edition. *TerraWatch Space*. 28 August.

There has been significant activity also in the dissemination layer in the past few years. Driving factors in this layer include the following:

- Accessibility (i.e., do not let the EO data remain hidden in the cloud).
- Interoperability (i.e., do not let the data be buried in application-specific vertical silos).
- Fusibility (i.e., allows easier and effective data integration wherein the whole is greater than the sum of its parts).
- Usability (i.e., prepare data as "analysis-ready" for the user, saving time and facilitating immediate processing).

The dissemination layer can be broadly categorized into three segments. A mapping of the main industry players within these segments is shown in Figure 5 (although these activities may overlap).

Figure 5: Mapping of Main Earth Observation Companies for Data Dissemination Based on Their Primary Function

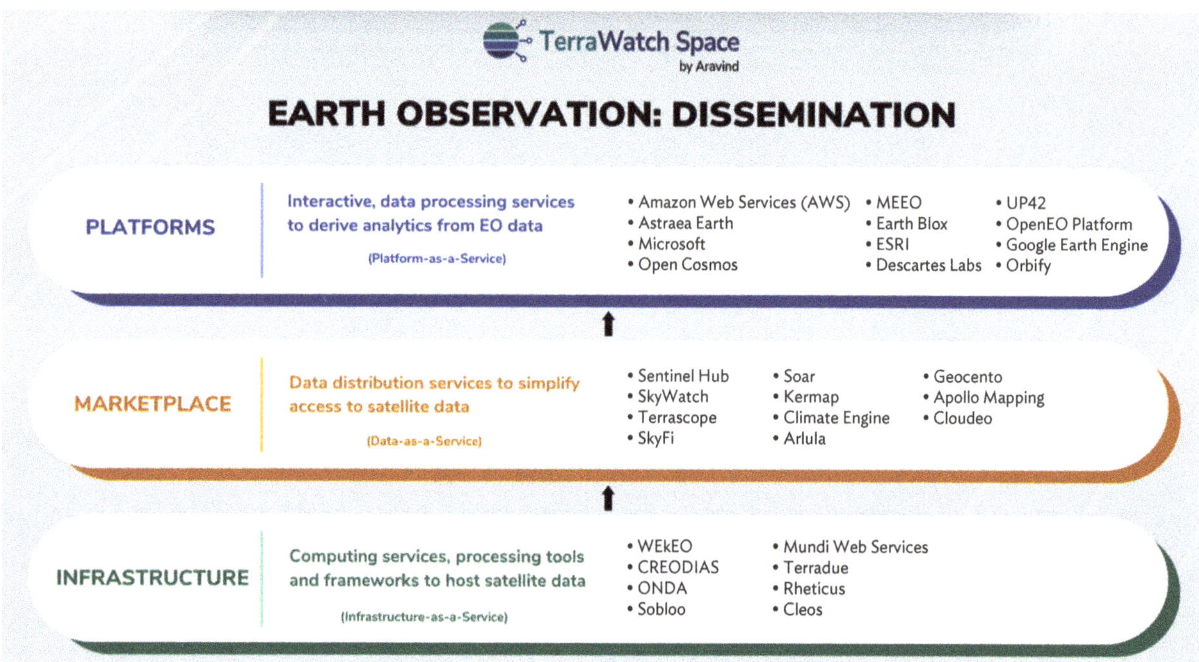

Source: Aravind. 2022. The State of Commercial Earth Observation: 2022 Edition. *TerraWatch Space*. 28 August.

The intelligence layer comprises a vast and highly diverse range of public sector organizations and rapidly growing private sector companies. Details are discussed in Section 2.4.3.

2.3 NewSpace Sector

Over the past decade, a disruptive and highly innovative trend known as "NewSpace" has emerged, characterized by the proliferation of private companies and small startups entering the space industry with radically different approaches, technologies, and business models. This new trend began and is well advanced in the US, with Europe now catching up. However, Asia and the Pacific is now poised to be the fastest-growing region in this innovative NewSpace technology sector over the next decade.

▲ Image from European Space Agency.

NewSpace is affecting all aspects of space technologies (e.g., reusable and low-cost launchers, manned space vehicles and exploration, and space-based global internet provision), but here, we focus on the aspects pertaining to EO.

The NewSpace revolution is being achieved through the use of standard, commercial off-the-shelf components for small satellite (SmallSat) electronics and platform structure (such as those mass-produced for cell phones, automotive industry, and recreational drones), thereby reducing costs and delivery times. These factors have led to completely different satellite missions, with constellations of hundreds of nanosats operating together to significantly increase global coverage and frequency of observation of the Earth's surface.

The NewSpace sector is being driven by a wide range of private sector entrepreneurs, focused on commercial objectives, and with growing access to venture capital. Satellite data are usually not free and are available only at commercial rates, but data costs are rapidly falling. "As a service" (AaS) business models are appearing across the value chain, including satellites, ground segments, and data analytics.

Emerging applications of NewSpace technology are starting to influence our lives, businesses, and the way we perceive and understand our own planet. Such developments are now causing an explosion of business opportunities (and some failures).

2.3.1 Key Drivers of NewSpace

Some of the key drivers of NewSpace are as follows:

- **Emergence of small satellites.** One of the defining features of the "NewSpace" sector is the rise of SmallSats, particularly micro and nano satellites. These smaller, less complex, and more affordable satellites are leveraging miniaturization and are using standard, commercial off-the-shelf components, thereby reducing cost and delivery times. SmallSats have democratized access to space, allowing small startups and universities to develop and launch their own EO missions with reduced costs and faster deployment times to operations.
- **Constellation-based approach.** SmallSats are increasingly being deployed in large constellations, providing global coverage and high revisit rates, even at very high resolution of a few meters.

These constellations enable monitoring of large geographic areas, real-time data acquisition, and improved temporal resolution. Companies like Planet, Spire Global, Capella Space, and Zhuhai Orbita are building large constellations capable of capturing EO data daily, revolutionizing the way we monitor and understand our planet.

- **Private sector financing.** Today, there are more ways of raising capital than ever before. The financing solutions are numerous—from business angels to venture capitalists, financial corporations, or the public markets. There has been a notable increase in investment activity in space companies within public markets over the last 3–4 years. ARKX was launched in March 2021.[58] It is an actively managed exchange traded fund that aims to achieve long-term growth of capital by investing in companies that are engaged in space exploration and innovation (82% of which are located in the US). These investors are focusing on NewSpace sector startups whose capabilities and initiatives could drive innovation and growth over the next several years. There has been a significant rise in interest over the last 2 years in special purpose acquisition companies (SPACs).[59] As a new means of financing, SPACs have enabled startups to enter the market and consolidate their assets to become profitable. The SPAC boom has reshaped the EO sector and is likely to grow further in the future.
- **Taking risks.** The traditional approach in EO has been a zero-risk development focused on technological reliability and performance with a number of payload instruments on-board a single, large satellite. The main satellite development programs have been implemented with tried-and-tested big aerospace companies with a well-established space heritage. This method has subsequently required extensive, exhaustive (and expensive) on-ground satellite assembly, integration, and test activities to minimize the risk of an "all-eggs-in-one-basket" type of failure. Through NewSpace, the development approach is much more accommodating of innovative technology. Constellations (consisting of hundreds of microsatellites) allow for multiple satellite in-orbit failure and therefore minimal on-ground testing. Each single satellite is cheap enough to fail and fast enough to replicate in the event of failure.
- **Data analytics and value-added services.** With the increasing volume of EO data being collected, companies in the NewSpace sector are revolutionizing data analytics and providing new value-added services through enhanced computing facilities and data processing technologies (platforms). By combining satellite imagery with AI, machine learning algorithms, and geospatial analytics, these companies can extract meaningful insights and actionable information for their customers. These value-added services not only enable information mapping but are increasingly handling large time-series of EO data to establish ease of monitoring and to detect changes while looking for correlations across data that were previously not possible. The applications cover a wide range of environmental domains, including land cover classification, vegetation monitoring, agriculture, urban infrastructure, biodiversity, change detection, and disaster responses.

2.3.2 SmallSat Constellations and NewSpace Entrepreneurs

The key driver of SmallSat constellations and the innovative data analytics that NewSpace entrepreneurs are bringing are explored in further detail below.

The satellite EO industry sector is highly competitive, with many players in the market. One of the main industry trends fostering market expansion is the rising demand for small satellites (SmallSats).

[58] Ark Funds.

[59] Matt Christie. 2021. *The Rise of SPACs.*

SmallSats can range in size from that of a shoebox to that of a washing machine and weigh as little as 5–10 kilograms (kg).

Euroconsult has identified more than 65 companies that have announced intentions to develop SmallSat constellations, which represent nearly 2,300 SmallSats (lighter than 500 kg), with a large majority under 50 kg.[60] Of approximately 65 companies, roughly 25% have claimed to deploy constellations of 50+ satellites, the majority having fewer than 30 satellites. However, not all the listed companies will reach the deployment phase because many of them will face funding issues.

Figure 6: New Commercial Earth Observation Constellation Companies and Operators

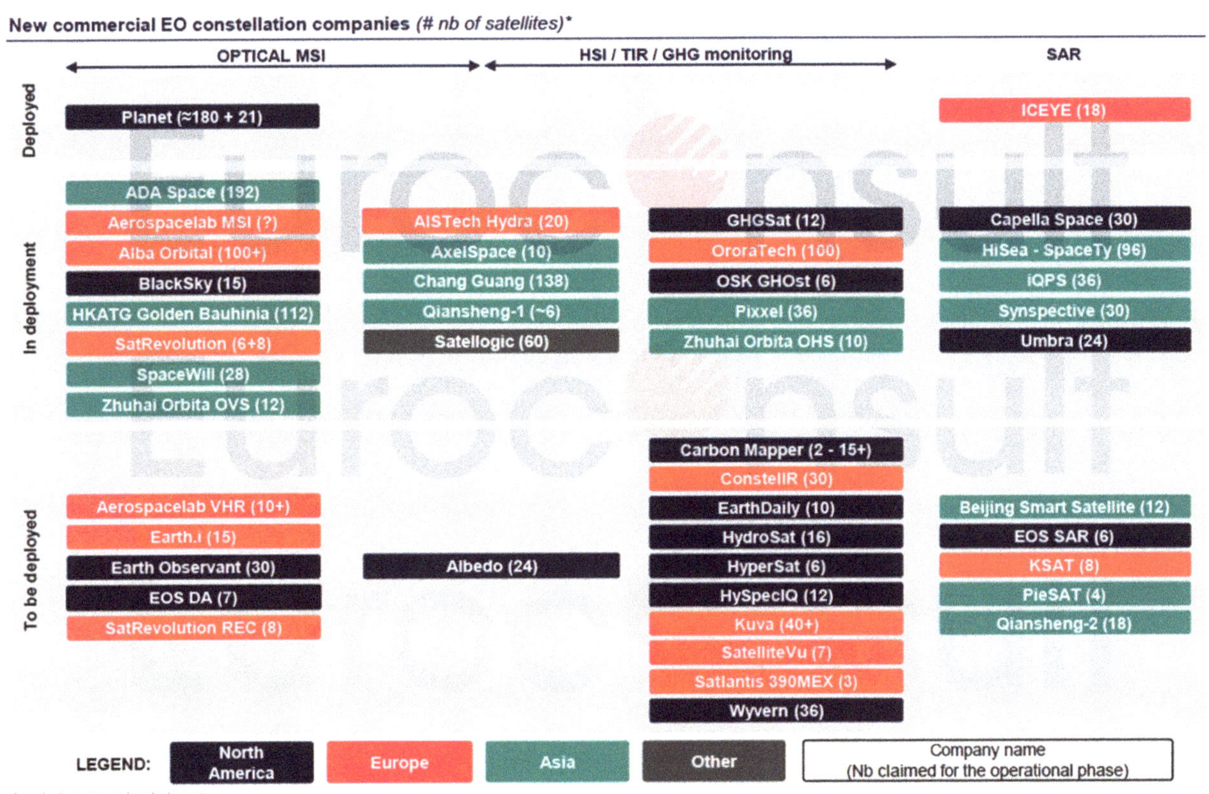

Source: Novaspace. 2022. Global market for commercial Earth Observation data and services to reach $7.9 billion by 2031. Press release. 16 December.

New constellations may offer different services, but they all focus on one theme: the ability to offer new services and better meet market needs based on high-frequency change detection analytics.

60 Euroconsult. 2024. *Earth Observation Data and Services Market*. 17th Edition.

In summary, satellite EO data have key attributes that make it a unique and valuable source of the Earth's environmental information. These attributes include

- information objectivity and known accuracies;
- data collection repeatability and comparability over time;
- global geographic coverage, including difficult to access regions;
- long-term data continuity to 2030 and beyond;
- data affordability, including free and open access to data;
- thematic information content expanding in detail and scope; and
- ever-increasing frequency of observations, including daily imaging capabilities.

Satellite EO provides a wide range of key thematic types of environmental information that is *global, comprehensive, accurate, repeatable, and timely*.

Satellite EO remains the only means to provide comprehensive, consistent, and up-to-date imagery of the Earth's global surface and of the changes taking place on that surface over time and space.

2.4 Data Access and Processing

To understand the various activities involved in EO data access, processing, and use, a generic concept of the EO information value chain is presented below.

Figure 7: Earth Observation Value Chain

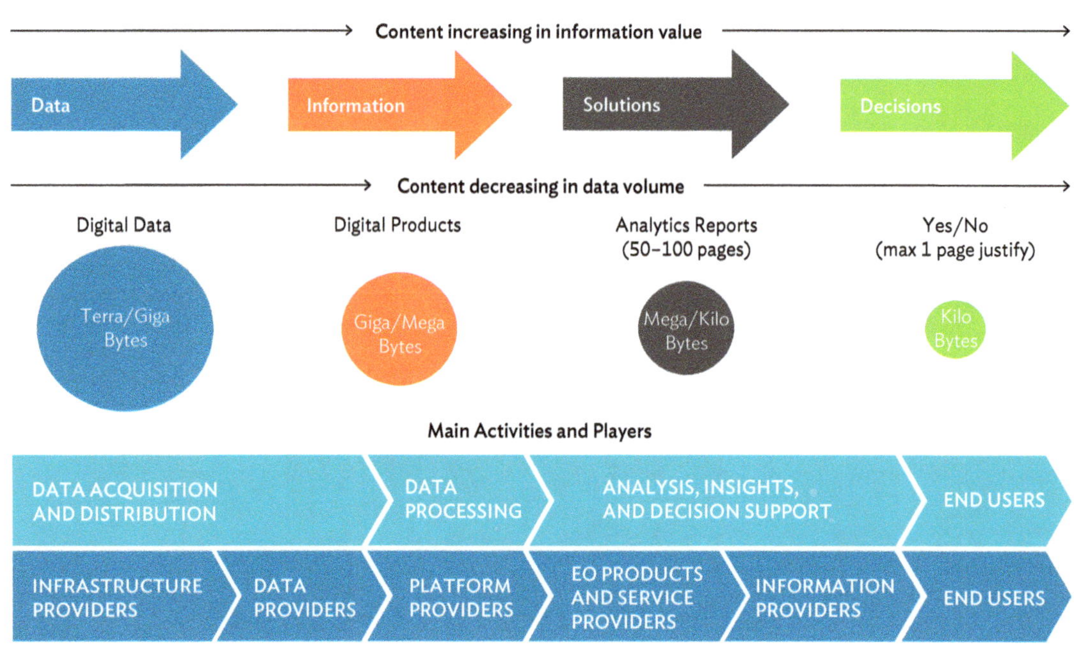

Source: EUSPA. *GNSS and EO Market Report*. p. 13.

The EO information value chain is depicted from four viewpoints of what is happening along the chain:

(1) Information content that transforms from basic satellite data to environmental information products, to solutions to environmental issues, to decision-making material (that leads to actions).

(2) Data volumes that are decreasing along the chain from typically terabytes and gigabytes of data at the start to a few kilobytes at the end (leading to a yes/no decision on what action is to be taken).

(3) Processes involved in transforming the data (i.e., the main steps in adding value to the data along the chain).

(4) Types of actors and communities involved in the processes along the chain.

Satellite manufacturers have a long history of moving downward to operate satellites, deliver data, and ultimately service to create growth opportunities. More recently, the opposite trend has been occurring with constellation players, who are vertically integrated with their proprietary manufacturing capability, moving up by extending their business to space systems supply, especially targeting satellite-as-a-service to complement revenues.

New service profiles are also joining the value-adding process. Cloud operators, aside from offering their storage services, are processing the first level of correction and are looking for more service virtualization processes. Historical software providers, such as ESRI (with ArcGIS online) or NV5 (which acquired L3Harris' geospatial software business as well as Axim Geospatial), are now selling online tools and processing chains which provide automation that are creating value-adding services (VAS).

A useful guide to help EO data providers and development practitioners within the EO value chain (from data to applications) to greatly improve how EO data is used in decision-making can be found in the paper, *Ten Rules to Increase the Societal Value of Earth Observations*.[61]

Table 1: Ten Rules to Increase the Societal Value of Satellite Earth Observations

Rule Action	Purpose	Approach
1. Identify the root causes of a problem	A well-defined problem helps direct problem-solving efforts toward addressing underlying issues rather than just its symptoms.	A good approach for root cause analysis is the "5 Whys?" method developed as part of the Toyota production system in 1988. It entails repeating the question "But Why?" five times as an iterative interrogative technique used to explore the cause-and-effect relationships underlying a particular problem and to identify its solution.[a]
2. Consider how data is really used	Understanding how different people need and want to use data and information can help identify the best solution.	Different approaches are available to understand users' needs and worldviews associated with data use. A stakeholder analysis is useful for identifying people's behavior, intentions, interrelations, agendas, and interests. User experience research is often used by technology companies to understand how users experience their products, so that their perspectives may be integrated into the design and functionality of their products.

continued on next page

[74] A. Virapongse et al. 2020. Ten rules to increase the societal value of earth observations. *Springer Nature Link*. 3 March.

Table 1 *continued*

Rule Action	Purpose	Approach
3. Get the right people to the table	Building bridges between data and use is not easy, but identifying and accessing key intermediaries and users can help.	Co-production of knowledge and co-implementation of projects are approaches that help to reach a high level of success for solutions. But it is not common that all of the identified people and groups will be easy to access and work with as different groups use different vocabularies and meanings, so it is essential to identify intermediaries. Intermediaries are knowledgeable and trusted in multiple cultures; they bring benefits like opening paths of communication, providing cultural translation, aligning interests, establishing trust between groups, and bringing the problem solving to a more concise level.
4. Investigate all alternative solutions carefully	Comprehensive identification and analysis of potential solutions helps determine which (if any) solutions should be advanced.	Before pursuing specific solutions to a problem, the breadth of different options must be identified and explored through ideation. Ideation is often conducted through different variations of brainstorming, which aim to inspire creative problem-solving by encouraging people to share ideas while withholding criticism or judgment. The group should consider what they like and dislike about each idea, as well as its potential negative and positive side effects, practicality, the potential effectiveness, and how easy or difficult it is to put into practice. Cost-and-benefit valuation methods should clearly link the use of data and methods to defined and quantifiable outcomes.
5. Evaluate, adapt, and iterate	Solutions must be strategically evaluated and refined to ensure their best possible fit within the context.	An evaluation strategy requires identifying the societal outcomes that a solution seeks to influence, metrics for tracking changes in those outcomes, and an empirical strategy to assess whether the changes in outcomes (as monitored by the metrics) can be attributed to the solution. Standard methodologies can be employed for monitoring and evaluation (M&E), such as the project's theory of change.[b]
6. Think globally	Data and data products should adhere to existing best practices, standards, and ethical considerations.	Encoding geospatial information in unified data standards is crucial to ensuring data interoperability, consistency, accuracy, integration, and sharing. It enables efficient data management, analysis, and collaboration, while supporting long-term data governance and facilitating integration across various systems and technologies.

continued on next page

Table 1 *continued*

Rule Action	Purpose	Approach
7. Trust is essential	Users must be able to easily determine that data and data products have been developed with transparency and scientific rigor.	A user's confidence and trust in data and information are greatly affected by their understanding and perceptions of the uncertainties in data, analyses, and products. The development of indices is one approach. Operational Readiness Labels (ORLs), for example, have been developed to aid users in planning, decision support, and risk reduction by creating a standard by which data may be evaluated and ranked for use by stakeholders in a uniform model. They range from the highest level of ORL 1 (data are available now for immediate decision-making and there are people available to contact with questions about it), to the lowest level of ORL 4 (data are still being evaluated for accuracy and being validated).[c]
8. Lower barriers to entry	Adapting outcomes to user capabilities can help increase the uptake and success of data and information solutions.	When users fail to appreciate the value of EO information, it is often because they struggle with information that they are uncertain of how to use. Both tailoring EO information products to align with user capability and building capacity among users to apply these products is essential. Working within user capability entails developing, translating, and communicating products that align as closely as possible to a user's context, such as their existing workflows, habits, language, and culture. Common capacity-building tools and mechanisms include guiding documents, webinar tutorials, workshops, and one-to-one expert support.
9. Document the process	Capturing lessons learned allows others to avoid repeating mistakes and to improve upon successes.	Documenting the project allows for impediments to knowledge transfer and uptake to be highlighted, as these subtle barriers to the application of EO data and products are easy to overlook. The goal of documentation is to create a record of presentable and useful information, not just create documents. Therefore, the project should plan early on what should be documented, how and when documentation should occur, how much documentation is needed, who should do it, and what the goals of the documentation are.
10. Walk away when the solution has legs	Once the solution has been adopted, know when it is time to let users take over.	To assess when it is time to allow users to take the lead, the project's theory of change can provide a helpful framework. Articulating the outcomes of a solution allows for monitoring and evaluation of progress to occur, so that when outcomes have been met, it can be identified that it is time to close the project and let the users take the lead.

[a] Businessmap. Unlock the Power of 5 Whys: Root Cause Analysis Made Easy.

[b] BetterEvaluation. Describe the theory of change.

[c] All Hazards Consortium (AHC). SISE.

Source: A. Virapongse et al. 2020. Ten Rules to Increase the Societal Value of Earth Observations. *Earth Science Informatics*. Vol. 13. pp. 233–247.

The 10 rules outlined here provide guidance for actions and best practices that data providers and intermediaries in the EO value chain can implement while planning and executing projects that seek to increase the societal value of EO data. In particular, the rules address the "value chain improvement" and "use-driven" approaches and focuses on meeting user needs, problem-solving within interdisciplinary teams, and long-term sustainable solutions.

As such, they are a pragmatic approach to helping both EO data providers and development practitioners within the EO value chain (from data to applications) in reducing the barriers to uptake of EO solutions.

2.4.1 Data Acquisition and Provision

The major players in operating EO missions in the institutional, government, private, and NewSpace sectors are presented in the previous three sections of this chapter. These mission operators cover a wide variety of EO capabilities and offer access to EO data under different conditions (i.e., data policy), ranging from free data access to access at cost with varying commercial rates.

The game changer in EO global data acquisition and access is the EU Copernicus system, with some 25 terabytes of data being collected each day (as of late 2023).

Figure 8: Evolution Over Time of Earth Observation Data Volumes for Missions Operated and Handled by the European Space Agency

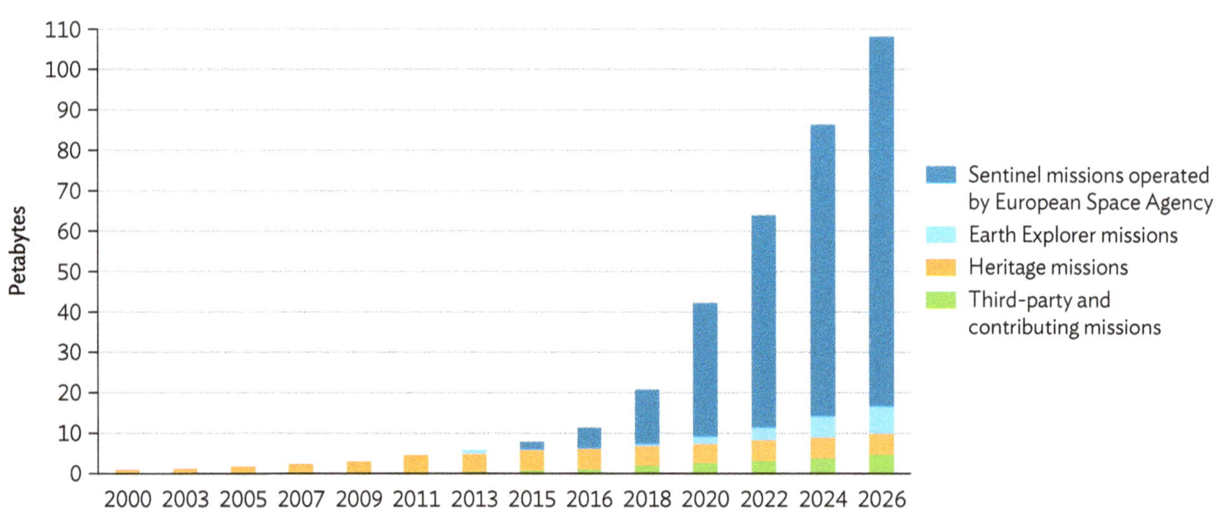

Source: European Space Agency.

Through the full, free, and open (FFO) data policy of the EU EO program, the data coming from the Sentinel satellites and the value-added products coming from the Copernicus services are made available free of charge to all users. Through the FFO, data and information services have been widely delivered across the globe and have formed an enabling basis for the development of further value-added products that have accelerated the commercial EO market. The Copernicus Data Space Ecosystem is constantly being upgraded to provide improved access to Sentinels' missions data.

Commercial EO missions offer EO data from a variety of optical multispectral, imaging radar, and hyperspectral instruments.

The EO data market stands at $1.78 billion (as of 2022), growing at a 5-year compound annual growth rate (CAGR) of 6%. Forecasts reveal further growth to reach a total value of more than $2.7 billion by 2032, at a 4% CAGR throughout the decade, which is mostly driven by the demand for very-high-resolution (submetric) imagery.[62]

Generally, for very-high-resolution EO data (i.e., submeter for optical and a few meters for radar), data costs are increasing. However, over the last decade, the EO data costs of commercial missions have been falling overall, a trend that is likely to continue.

2.4.2 Data Processing

With the advent of high-performance cloud computing infrastructures and data analysis techniques, complex environmental monitoring problems are beginning to be solved with EO-based information.

New and innovative geospatial data processing infrastructures and platforms are key to realizing the potential of EO, especially given the volume and complexity of EO data now being produced.

Some of the main processing infrastructures are described below.

Copernicus Data Space Ecosystem

Not only does this system provide access to Sentinels' mission data, but it also provides a powerful data analytics environment with access to a set of data processing tools that allow generation of higher-level, value-added information suitable in supporting scientific, public sector, and private sector or commercial activities.[63]

▲ Copernicus Data Space Ecosystem commercial service providers (image from the European Union).

The ecosystem allows users to process and analyze the data with cloud computing resources that can be directly activated in a personal user account. In addition, the openEO algorithm plaza can host algorithms that are built on top of the openEO applications programming interface (API). These algorithms can be accessed via user-friendly interfaces or via the API that is to be integrated into new services and/or platforms. Finally, the ecosystem provides commercial services to support cloud migration, offering a complete lifecycle for tailored EO applications and services, from design to operations, and including the strong security environment of IT platforms.

---

62 Novaspace. 2023. Global market for commercial Earth Observation data and services to reach $7.6 billion by 2032. Press release. 28 November.

63 Copernicus Data Space. Analyse.

EU WEkEO Environment

WEkEO is the EU Copernicus reference service for environmental data, virtual processing environments, and skilled user support.[64]

▲ WEkEO implementing organizations (image from the European Union).

WEkEO offers coherent and harmonized access to a wide range of varied EO data sources through a single protocol. It is built on a distributed infrastructure connected by a high-speed network, providing hosted cloud-based, on-demand processing. It gives access to historical and contemporary data from models, satellites, and in-situ sensors from the Copernicus atmosphere, climate, marine, and land services. The WEkEO Help Centre provides expert online support to users.

European Space Agency OpenEO Platform

The openEO platform provides a set of intuitive programming libraries to process a wide variety of EO datasets.[65] This allows different levels and types of EO data processing activity to be carried out, from exploratory scientific research to pre-operational large-scale production of EO-derived maps and information.

▲ Images from the European Space Agency (ESA).

The openEO platform can be used in a wide variety of programming languages and environments, including the Jupyter Lab for interactive prototyping, programming, and visualization, and support to python, JavaScript, and R-language. In addition, the Platform Editor provides a visual and interactive user interface in the browser, which allows monitoring of the status of processing workflows without requiring detailed programming experience.

Through the openEO platform, users can also apply for technical and financial development support for new EO applications through the ESA Network of Resource (NoR) initiative.[66] The NoR supports diverse scientific, educational, and pre-commercial usage of openEO platform via the sponsoring mechanism, which is free of charge. The aim is to support users to innovate their working practices, moving from a "data download" paradigm toward a "bring the user to the data" paradigm, a change considered to be essential given the hugely growing volumes of EO data now available.

---

64 WEkEO.

65 openEO Platform.

66 EO Science for Society. Network of Resources.

European Space Agency Thematic Exploitation Platforms

With the exponentially growing volume of basic EO data becoming available, it is a real challenge to ensure these complex data streams are exploited to their full potential for science and applications. ESA is addressing this challenge with Thematic Exploitation Platforms (TEPs).[67]

ESA has developed seven TEPs, each addressing a specific thematic domain (i.e., coastal, forestry, hydrology, geohazards, polar, urban, and food security).

Each TEP is a collaborative, virtual work environment that provides access through a single coherent interface to the thematically relevant EO and non-EO data sources, to computing resources, and to hosted processing (i.e., Infrastructure as a Service) and a platform environment (i.e., Platform as a Service).

The TEPs allow users to integrate, test, run, and manage new applications (i.e., EO data processing) in the six specific thematic domains without the complexity of building and maintaining their own infrastructure.

Figure 9: Seven European Space Agency Thematic Exploitation Platforms

Source: European Space Agency. Thematic Exploitation Platforms (videos).

EO Data Cubes

Data cubes are multidimensional arrays of stacked, EO image data, aggregated from a variety of sources but standardized and harmonized to be analysis ready (i.e., to be used in further analytics with minimal effort).[68] Data cubes are deployed on the cloud for efficient, on-demand data processing.

The production of analysis-ready data significantly reduces the burden of further EO data processing by minimizing the time and scientific expertise required to access and prepare varied EO data sources.

67 ESA. Thematic Exploitation Platforms.

68 Earth System Data Lab.

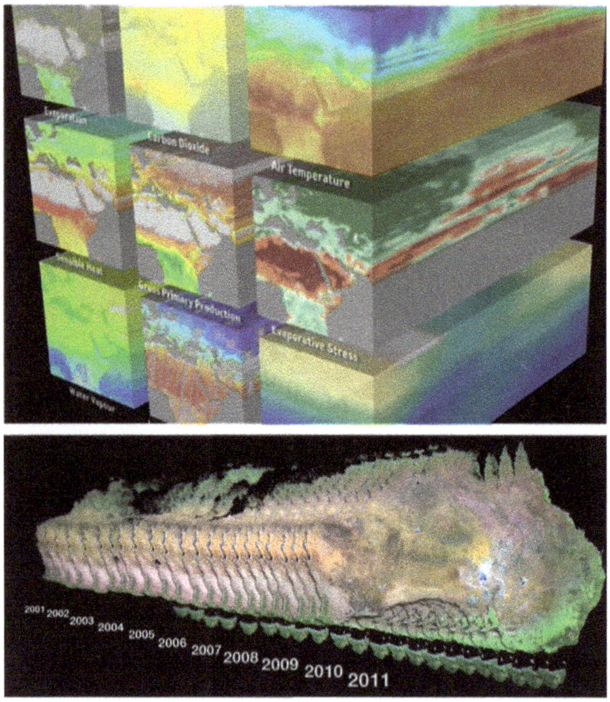

▲ Examples of Earth Observation data cubes: Earth in a Box (top) (image from Earth System Science Lab of European Space Agency; Time series of optical multispectral imagery for the continent of Australia (bottom) (image from Australian GeoScience Data Cube).

Figure 10: Results of a Survey on How Data Scientists Spend Their Time

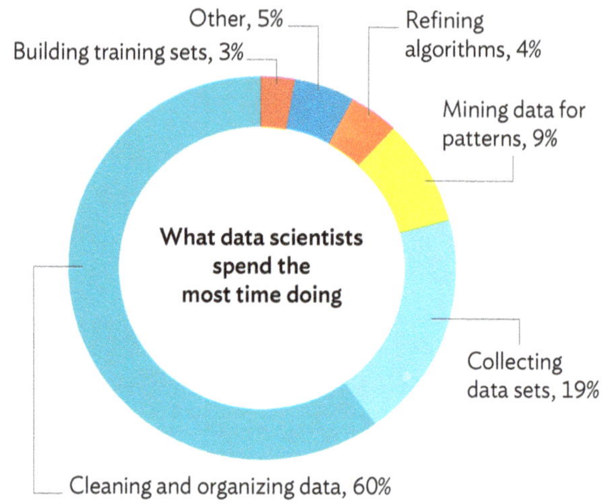

Source: G. Press. 2016. Cleaning Big Data: Most Time-Consuming, Least Enjoyable Data Science Task, Survey Says. *Forbes*. 23 March.

Data preparation accounts for about 80% of the work of EO scientists and 76% of them consider this arduous task as the least enjoyable part of their work.[69] The aim of providing analysis-ready data is a key goal of the data cube.

Brief descriptions of some of the main data cubes currently available are as follows:

Africa Regional Data Cube

The Africa Regional Data Cube (ARDC) was launched in May 2018. It is both an infrastructure and a service providing analysis-ready EO data vehicle in support of Sustainable Development Goals (SDGs).[70] ARDC has been developed and operated to support Ghana, Kenya, Senegal, Sierra Leone, and Tanzania in addressing national issues relating to agriculture, food security, deforestation, urbanization, and water access.

In 2020, ARDC transitioned into Digital Earth Africa.[71] Working in close collaboration with the AfriGEO community, Digital Earth Africa is driven by environmental information needs, and challenges and priorities of the African continent. Digital Earth Africa will build on existing EO capacity and leverage these resources to address key issues across the continent with decision-ready EO information products for immediate use.

[69] G. Press. 2016. Cleaning Big Data: Most Time-Consuming, Least Enjoyable Data Science Task, Survey Says. *Forbes*. 23 March.

[70] Global Partnership for Sustainable Development Data. Africa Regional Data Cube.

[71] Digital Earth Africa.

Digital Earth Africa is hosted within Africa and has its own independent governance framework. The governing board includes organizations such as UN Economic Commission for Africa, African Union Commission, African Development Bank, Group on Earth Observations, and World Economic Forum.

The Technical Advisory Committee includes technical institutions such as Regional Centre for Mapping of Resources for Development, AGRHYMET, South Africa National Space Agency, African Regional Institute for Geospatial Information Science and Technology, Critical Earth Observations, and Ghana National Statistical Office (to name a few).

▲ Graphics from Africa Digital Regional Data Cube.

Digital Earth Pacific

Digital Earth Pacific is an operational earth and marine observation system that provides easier access to decades of satellite EO data.[72] It supports decision-making across the Pacific region through the Pacific Community (SPC), an international development organization owned and governed by 27 countries and territory members. SPC has been active in regional development since 1947.[73]

Digital Earth Pacific has Microsoft's Planetary Computer at its core, which provides free, open, and ready-to-use EO products and services as a public good investment. Digital Earth Pacific forms an underlying digital infrastructure that will ensure every nation in the Pacific has access to tools, technologies, and processing capacity to routinely monitor and track local environmental challenges, such as climate variability, food insecurity, and natural hazard events, as well as progress toward the achievement of SDGs.

▲ Key focus areas of Digital Earth Pacific (image from Digital Earth Pacific).

72 Digital Earth Pacific.

73 Pacific Community (SPC).

Euro Data Cube

The Euro Data Cube offers a set of commercial EO services to access and analyze a range of EO data in one application.[74]

ADB Cloud-Based Satellite Earth Observation System (Cloud SEOS)

The ADB Cloud SEOS is a prototype cloud-based platform currently under development.[75] It supports the use of machine learning techniques to allow key stakeholders to process free and open satellite data from sensors at different spatial resolutions. It is an initiative of the former Digital Technology for Development Division of the Climate Change and Sustainable Development Department, now the Digital Sector Office in the Sectors Department 2 and the Information Technology Department.

Satellite information mapping layers planned to be available include land cover, coffee, mangroves, water quality, coastlines, vegetation indexes, crop types, tree condition, tree crop types, tree crop yield estimation, food crop condition, food crop biomass, food crop yield estimation, and climate indices.

The aim of Cloud SEOS is to host AI models and satellite-based imagery, to be made available through open and cloud-based platforms. Capacity building will be facilitated to foster uptake in developing member countries that lack EO skills and major investments in local IT infrastructure. This prototype is currently used to design investment and policy projects in Southeast Asia.

Large ICT Players

Google Earth Engine is a computing platform that enables users to execute diverse geospatial analyses on Google's cloud infrastructure.[76]

▼ Images from Google.

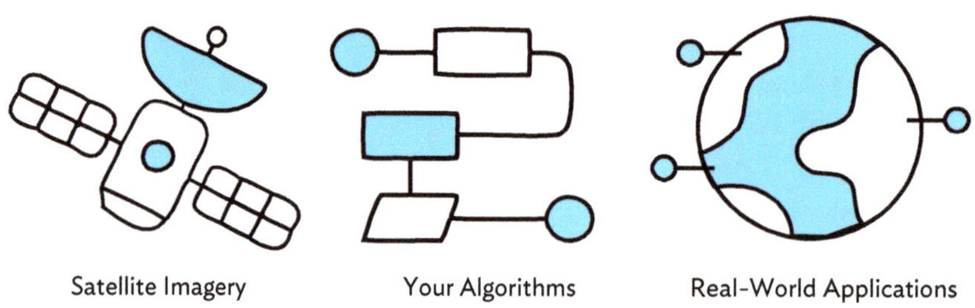

[74] Euro Data Cube.

[75] ESA Global Development Assistance (GDA). 2024. ESA and ADB taking effective next steps on adopting EO for Asia-Pacific development. 25 September.

[76] Google Earth Engine.

Google Earth Engine combines a huge (multi-petabyte) catalogue of satellite EO imagery and geospatial datasets with global-scale analysis capabilities. Earth Engine's public EO data archive includes more than 40 years of historical satellite imagery (including LandSat, MODIS, and Sentinels imagery) and scientific datasets, which are updated and expanded daily. This large, historical, and global EO data archive, together with flexible data analytics tools, allows users to readily use Earth Engine to detect changes and map trends for many physical aspects of the Earth's surface. Google Earth Engine is made available at a cost for commercial use, but is offered free for academic and research use.

Amazon Web Services (Earth on AWS) allows users to build global-scale applications in the cloud with open geospatial data.[77]

The registry of open data on AWS supports users in discovering and accessing EO datasets that are available via AWS resources. AWS cloud credits are available for the scientific community to conduct research using EO data on AWS. This community is the primary source of technological innovation and is supported to make novel developments in the EO applications sector.

There are many use case videos and articles explaining how EO data on AWS can be used in public sector activities, business enterprises and new startups, and research institutions.

Microsoft's Planetary Computer combines a multi-petabyte catalogue of global environmental data with intuitive APIs, a flexible analytics environment that allows users to explore global issues about that data, and applications that provide answers to a range of environmental stakeholders.[78]

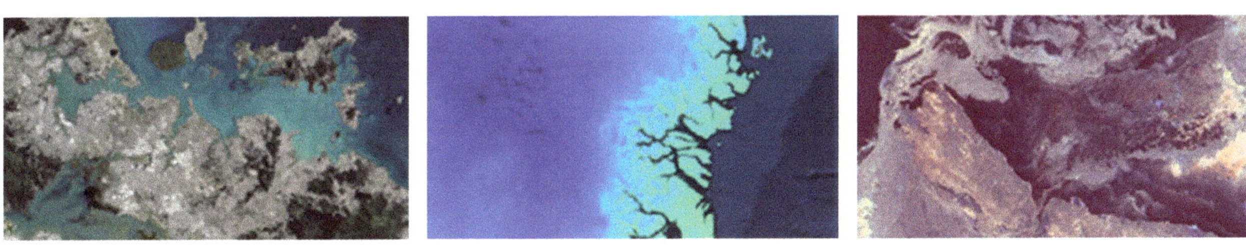

▲ Images from Microsoft.

The satellite data archive includes more than 40 years of historical satellite imagery (including LandSat, MODIS, and Sentinel-1 imagery). As part of ArcGIS for Planetary Computer, ESRI is bringing virtual ArcGIS Pro machines to the cloud, providing access to hundreds of ready-to-use analysis tools with an intuitive user experience. From interactive suitability modeling to AI for predictive analytics, these preconfigured virtual machines exploit the expansive collection of analysis-ready data in Planetary Computer and capture actionable insights.

[77] Amazon Web Services. Earth on AWS.

[78] Planetary Computer.

AI and Machine Learning Technologies

As described earlier in this chapter, the volume of EO data being generated is growing exponentially and exceeds the current capacity to analyze all of it. Using AI and machine learning techniques can help decipher these data, as computers have the capability to analyze vast volumes of data to identify patterns and correlations not possible to see with conventional analytics approaches.

The World Economic Forum (WEF), in collaboration with Deloitte, produced a briefing paper, *The Catalytic Potential of Artificial Intelligence for Earth Observation*,[79] that provides vital insights for taking WEF's EO work forward. This work supports organizations in contributing to a nature-positive and net-zero economy, in setting science-based decarbonization targets, and in understanding their dependencies and impacts on nature.

According to the WEF Annual Meeting held in Davos in January 2024, the use of AI, coupled with low-cost, high-performance computing applied to EO data together with data from GPS-enabled and Internet of Things devices, will allow the following:[80]

- Answering complex questions. AI capabilities allow for more data to be processed quickly and accurately, enabling the transformation of vast reams of raw EO measurements into actionable insights.
- Providing intuitive user interfaces for non-expert users. In the same way that ChatGPT has made large language models ubiquitous, the development of more intuitive user interfaces will enable not only data scientists but also business users to access and use AI-enabled EO insights.
- Driving business model innovation. As AI makes EO more accessible, organizations across sectors and industries can scale up its application, disrupting commercial and sustainability-focused business models alike.

AI and machine learning techniques offer the potential of providing actionable insights and business cases for a range of applications, such as monitoring remote infrastructure, assessing climate risk, and adaptively managing supply chains.

Digital Twins

A digital twin is a virtual model of a physical object. A digital twin works by digitally replicating a physical asset in the virtual environment, including its functionality, features, and behavior.

Digital twins can replicate many real-world items, from single pieces of equipment to full, complex systems. Digital twin technology allows us to explore and test the performance of an object, to identify possible misfunctions, and to make better-informed plans for the object's maintenance and operations lifecycle.[81]

[79] World Economic Forum. 2024. The Catalytic Potential of Artificial Intelligence for Earth Observation.

[80] M. Singh. 2024. How AI and Earth observation can help life on our planet. *World Economic Forum*. 31 January.

[81] AWS. What is Digital Twin Technology?

Digital Twin Earth

Examples of Digital Twin Earth developments that will help visualize, monitor, and forecast natural and human activity on the planet are as follows:

Destination Earth (DestinE) is a flagship initiative of the European Commission to assist users in designing accurate and actionable climate variability adaptation strategies and mitigation measures via the development and use of a highly accurate digital model of the Earth system.[82]

Destination Earth

European Commission

DestinE will support efforts in addressing complex environmental issues through the following:

- Model and monitor the entire Earth system (land, marine, atmosphere, and biosphere) and assess the interaction and impact on the planet system of human interventions.
- Plan for rescue and recovery activities in response to potential environmental hazard events in order to mitigate socioeconomic crises, save lives, and avoid large economic impacts.
- Develop and test a range of actionable scenarios for improving sustainable development.

DestinE is an essential pillar of the European Commission's efforts toward the Green Deal and Digital Strategy. The development of DestinE was launched in 2022 with the goal of reaching full operational capability in 2030.

NASA's Earth System Digital Twin (ESDT) is developing new technologies for integrating diverse Earth system and human activity models with complete and continuous EO data and informational capabilities.[83] The goal is to provide Earth system information and predictions that can be used for better environmental management as well as for planning and supporting decision-making.

The three major components of an ESDT are a constantly updated digital replica of the Earth system, a set of dynamic forecasting models, and a wide range of tools and capabilities to assess impacts.

On 26–28 October 2022, NASA's Earth Science Technology Office co-organized with the Earth Science Information Partners the 2022 ESDT Workshop and produced a detailed report on ESDT systems, prototypes, technologies, and frameworks.[84]

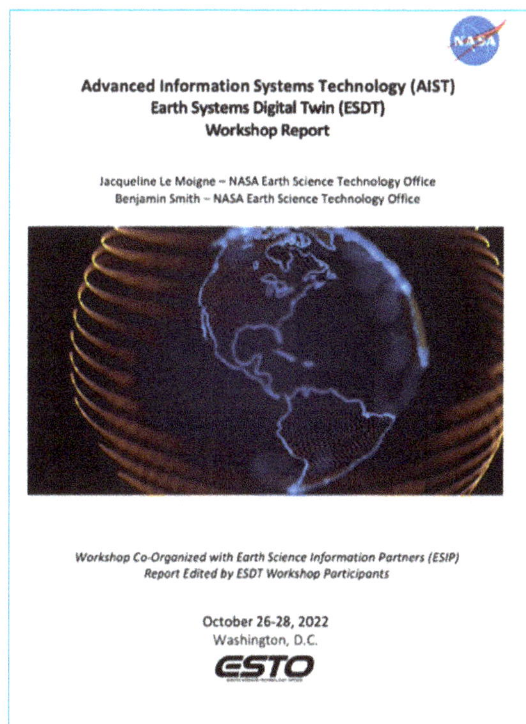

Advanced Information Systems Technology (AIST)
Earth Systems Digital Twin (ESDT)
Workshop Report

Jacqueline Le Moigne – NASA Earth Science Technology Office
Benjamin Smith – NASA Earth Science Technology Office

Workshop Co-Organized with Earth Science Information Partners (ESIP)
Report Edited by ESDT Workshop Participants

October 26-28, 2022
Washington, D.C.
ESTO

[82] Destination Earth.

[83] NASA's Earth Science Technology Office. Earth System Digital Twin.

[84] J. Le Moigne and B. Smith. 2022. *Earth Systems Digital Twin (ESDT) Workshop Report.*

Digital Twin Ocean

The European Digital Twin Ocean (DTO) was launched in February 2022.[85] It aims to make ocean-related information widely available to citizens, entrepreneurs, scientists, and policymakers through an innovative set of user-focused, interactive visualization tools.

Leveraging on existing European science and assets, the European DTO will provide a set of coherent and consistent high-resolution, multidimensional model descriptions of the oceanic system. This includes its physical, chemical, biological, socio-ecological, and economical dimensions, with forecasting timescales ranging from years to decades.

DTO will be fed by a stream of continuous real-time observations from thousands of in-situ sensors across the world's oceans and a range of EO satellites. The DTO will carry out advanced modeling using AI and machine learning techniques via supercomputing applied to historical EO data archives. The insights and knowledge generated by DTO will support better measures to mitigate and adapt to climate variability, to manage and preserve marine and coastal habitats, and to sustainably grow the blue economy markets.

2.4.3 Information Services for Decision-Making

Value-Adding Services and Analytics

EO value-adding services (VAS) are generated and provided by a vast and highly diverse range of public sector organizations and rapidly growing private sector companies.

For public sector organizations, examples include the following:

- **Asia and the Pacific.** The Asian Institute of Technology Geoinformatics Center uses AI and machine learning in analyzing satellite imagery, geographic information systems, and remote sensing data.[86] This has led to new discoveries and improved decision-making in a range of fields, including environmental science, public health, urban planning, and disaster management. The Indian Space Research Organization (ISRO) uses Space-Based Earth Observation Applications for agriculture, forestry, mining, water management, and rural and urban development.[87] The Korea Aerospace Research Institute (KARI) operates the National Satellite Information Support Centre and provides the information generated from the national Arirang and Cheollian satellites for various fields such as land management, disaster monitoring, marine water resource management, agricultural and forestry applications, environmental weather observation, national security, industrial activities, and daily life.

85 Mercator Ocean International. European Digital Twin of the Ocean.

86 Asian Institute of Technology. Geoinformatics Center.

87 ISRO, Department of Space. Space Based Earth Observation Applications.

- **Europe.** The European Commission Knowledge Centre on Earth Observation is providing systematic monitoring of EU policy requirements and priorities for Copernicus products and services.[88] It then formulates solutions to needs by applying state-of-the-art science to produce policy-tailored environmental information services for the European Green Deal and the Digital Agenda.

- **United States.** The NASA Earth Observations is available for browsing and downloading satellite imagery from NASA's constellation of Earth Observing System satellites.[89] This includes more than 50 different global datasets represented with daily, weekly, and monthly updates; to help capture climate-related and environmental changes as they occur on our planet.

EO information services have traditionally been provided by highly specialized small and medium-sized enterprises, operating usually in niche market sectors, starting in the early 1980s. Many of these companies originated from the academic R&D sector before spinning off into commercial startups. Due to this high level of technical diversity and fragmentation, the capabilities and products of this specialized EO services industrial sector are generally not easily accessible and understandable to the external user community.

However, in Europe, this has changed with the formation of a body representing this sector, the European Association of Remote Sensing Companies (EARSC).[90]

EARSC is a membership-based, not-for-profit organization which coordinates and promotes the activities of European companies that develop, produce, and deliver EO-derived data and environmental information.

Member companies operate across the full EO value chain—from data acquisition through processing, fusion, analytics, and provision of final geo-information products and services. EARSC represents this sector in its broadest sense, creating a network between industry, development organizations, decision-makers, and users.

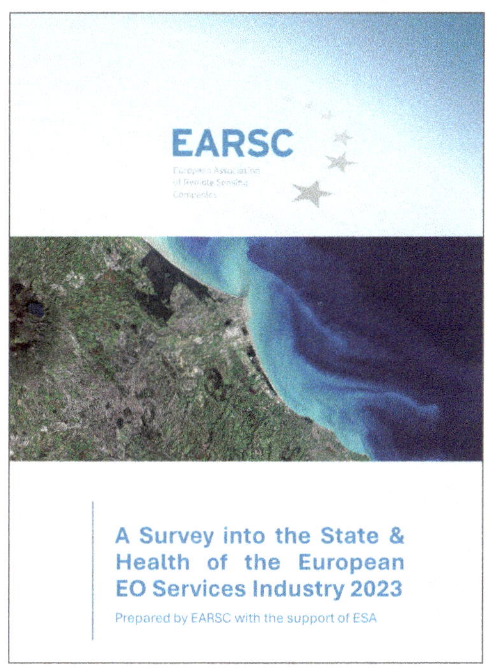

88 European Commission. Earth Observation.

89 NASA Earth Observations.

90 European Association of Remote Sensing Companies (EARSC).

In addition, since 2013, EARSC has carried out an annual, in-depth survey of the European EO services sector, highlighting key industry metrics, such as the number and distribution of companies, employment figures, and revenue trends, all of which offer invaluable insights into the European EO landscape.[91] This comprehensive annual report gives us a good understanding of technical geo-information services available and the factors driving the evolution and growth of the sector.

Figure 11: European Earth Observation Information Services Industry At a Glance (2023 Data)

Note: This figure is taken from EARSC Industry survey published in 2024, but using the industry data associated with 2023.

Source: European Association of Remote Sensing Companies (EARSC). Industry Facts & Figures.

The European EO VAS industry is well-developed in terms of technical capability due to a long history of public investment, both at the European level (EU, ESA programs) and at national levels.

Table 2: European Earth Observation Services Industry Market Shares for the Analytics Market Segments, 2019

	Europe's 2019 market share in analysis, insights, and decision support, by market segments						
	Analysis, insights, and decision support (Europe: 50%)						
	Europe	North America	Asia+Russia		Europe	North America	Asia+Russia
	64%	33%	3%		59%	31%	11%
	45%	44%	11%		93%	4%	1%
	60%	28%	12%		35%	16%	49%
	81%	8%	11%		45%	37%	18%
	89%	11%	0%		43%	28%	28%
	42%	31%	27%		17%	31%	52%
	47%	53%	0%		12%	77%	10%

Note: Segment share for "Rest of the World" is not shown in this table.

Source: European Space Programmes Agency (EUSPA). 2022. *EUSPA EO and GNSS Market Report*. 2022 edition, p. 14.

[91] EARSC. Industry Facts & Figures.

European companies accounted for half of the global EO analysis, insights, and decision support in 2019. The European industry is the market leader in most thematic segments within this analytics category. In each of the sectors of maritime and inland waterways, fisheries and aquaculture, and aviation and drones, the European industry makes up for more than 80% of the global market.[92]

In the EO data market in 2021, the North American companies led with around 55% of the market. This large share is due to the presence of very large companies including AWS, Alphabet (Google), and Maxar. Europe accounted for around 35% of the market in data processing and had significant market shares in the aviation, climate, environment, biodiversity, and insurance and finance segments.[93]

For developing countries, London Economics Europe's analysis for the EUSPA Market Report in 2022 estimated that a total of 241 companies based in developing countries were active in the EO industry in 2019, based on a bottom-up analysis of individual companies.

These companies contributed a total of €82 million to the global EO market in 2019 (2.2% of the global market). Given that the revenues were approximately €32 million in 2012, this implies that the industry in developing countries grew at a compound annual growth rate (CAGR) of 14%, which is greater than the 9% CAGR for the global EO services industry over the same period. This industry is estimated to have comprised a total of 1,250 employees in EO-related companies, or 7.8% of total EO services global employment for 2019.[94]

The relatively low contribution of developing countries to the global EO market is hypothesized as emanating from a lack of affordability and accessibility of EO data due to unreliable access to the internet. These factors lead to a scarcity of the specific technical skills required to produce EO information services. However, this skills gap could be readily filled by transferring the available EO skills in developed countries into those countries that are developing these EO capabilities. Such actions would open a significant opportunity for future growth in the digital economy of these developing countries.

Finally, with the advent of private sector SmallSat constellations in the last decade, most of these operators offer not only access to satellite imagery but also a wide range of user-oriented information services. As the primary purpose of these constellations is to generate business, these SmallSat operators generally have strong customer orientation and communication skills leading to easier access to the EO information services sector.

Global Information Services

In addition to VAS and analytics tailored to users' specific needs and providing targeted solutions to specific environmental issues, there is a wide range of (more standardized) global environmental information products available, many of which can be accessed via digital platforms and portals free of charge.

92 EUSPA. 2022. *GNSS and EO Market Report.*

93 EUSPA. 2024. *GNSS and EO Market Report.*

94 ESA GDA. 2022. *Wider Economic Benefits from Satellite EO in Developing Countries.*

Some examples of such global information products are as follows:

■ EU Copernicus Services

The Copernicus Services provide a range of information services in six thematic domains with up-to-date and accurate environmental information derived from the Copernicus Sentinels and cooperating missions satellite data, together with available in situ or open-data sources.[95]

Through the Copernicus program, long-term continuity and improvement of this services portfolio is guaranteed through 2030 and beyond.

A brief description of the types of environmental information included in these six thematic service domains is given in Table 3.

Figure 12: Copernicus Services Six Thematic Domains

Source: Images from the European Union (EU).

Table 3: Copernicus Services Content Description

	The **Copernicus Atmosphere Monitoring Service (CAMS)** continuously monitors the composition of the Earth's atmosphere at global and regional scales and delivers accurate and reliable information related to air pollution and health, solar energy, greenhouse gases and climate forcing. The European Centre for Medium-range Weather Forecasts (ECMWF) implements CAMS under the Delegation Agreement with the European Commission.
	The **Copernicus Marine Environment Monitoring Service (CMEMS)** provides regular and systematic reference information on the physical and biogeochemical state, variability, and dynamics of the ocean and marine ecosystems for the global ocean and the European regional seas. Mercator Ocean International was selected by the European Commission to implement CMEMS.
	The **Copernicus Land Monitoring Service (CLMS)** delivers geographical and environmental information on land cover, which includes land cover characteristics and changes, land use, vegetation state, water cycle, and Earth surface energy variables. The CLMS has been jointly implemented by the European Commission and the European Environment Agency.
	The **Copernicus Climate Change Service (C3S)** provides authoritative information about the past, present, and future climate, as well as tools to enable climate change mitigation and adaptation strategies by policymakers and businesses. C3S is being implemented by the ECMWF on behalf of the European Commission.
	The **Copernicus Emergency Management Service (CEMS)** delivers on-demand geospatial information* for emergency situations that arise from natural or human-made disasters anywhere in the world, and provides early warning information for floods, wildfires, and droughts. The CEMS has been implemented by the European Commission.
	The **Copernicus Security Service* (CSS)** delivers information which helps the European Union (EU) to improve crisis prevention, preparedness, and response, and to face the security-related challenges in the areas of maritime surveillance, border surveillance, and support to EU external action. The European Commission has entrusted the European Maritime Safety Agency (EMSA), European Border and Coast Guard Agency (FRONTEX), and EU SatCen with the different components of the CSS.

* = Services registered to Public users only.

Source: European Space Programmes Agency (EUSPA). 2022. *EUSPA EO and GNSS Market Report*. 2022 edition, p. 14.

95 Copernicus. Copernicus Services.

European Space Agency Climate Change Initiative

ESA's Climate Change Initiative (CCI) is a major R&D effort that generates global, decadal timeframe satellite data records to track and understand key scientific aspects of the Earth climate system.[96] This knowledge is key to the models and climate services used to inform and support climate action and report progress toward the Paris Agreement goals.

The Global Climate Observing System (GCOS) provides a framework for meeting the full range of national and international requirements for climate observations. GCOS has defined 55 Essential Climate Variables (ECVs) that are required to systematically characterize Earth's climate system (each ECV is a physical, chemical, or biological variable or group of linked variables). More than half of these ECVs can be measured from space.

The CCI program enables a community of more than 500 experts from across Europe to develop high-quality data products in a set of 27 research projects, with each project addressing and focusing on one of GCOS ECVs. Each ECV is validated against independent datasets and includes quantitative estimates of accuracies.

Figure 13: Examples of Global Satellite-Derived Climate Datasets from the European Space Agency's Climate Change Initiative

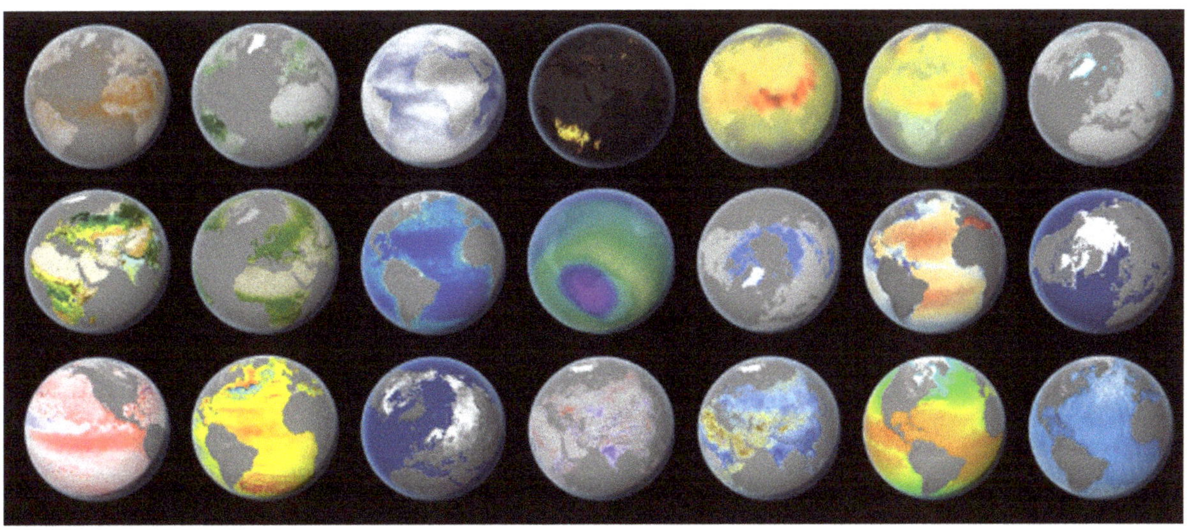

Source: Images from the European Space Agency.

NOAA National Centers for Environmental Information

The US National Oceanic and Atmospheric Administration's National Centers for Environmental Information (NCEI) provides a wide range of environmental data, products, and services.[97]

NOAA National Centers for Environmental Information
NATIONAL OCEANIC AND ATMOSPHERIC ADMINISTRATION

96 ESA Climate Office. Climate Change Initiative.

97 National Centers for Environmental Information (NCEI). Products.

In addition, NCEI hosts several world data centers and services through membership in the International Council for Science World Data System, which is a collaborative effort to ensure open access to environmental data around the world. NCEI has four national locations, as well as regional centers, field sites, and cooperative institutes, all spread throughout the US.

NCEI offers map viewers that contain data visualization tools that can display data from a variety of different sources in the same viewing environment, together with tools that can correlate this information with specific geographic locations.[98] The types of mapping information supported includes marine geology, geophysics, and bathymetry; ecosystems and natural resources; and natural hazards, disasters, and severe weather and climate monitoring.

2.5 International Initiatives

Collaboration between countries and international organizations in the EO domain has increased data accessibility and sharing, further enhancing the benefits of EO on a global scale.

A few examples of some of the main international organizations operating in the EO domain are as follows:

Group on Earth Observations (GEO)

GEO is an intergovernmental partnership working to improve the availability, access, and use of openly available EO satellite imagery, remote sensing, and in-situ data.[99] The overall aim is to promote the use of EO to impact policy and decision-making in a wide range of sectors. Formed in 2004, GEO now consists of 105 governments, 126 participating organizations, and extensive EO resources in the GEO Knowledge Hub.[100]

GEO's main priorities include the SDGs, climate action, and disaster risk reduction. GEO makes available EO-based information in support of policy formulation for climate variability adaptation and mitigation. GEO also supports international coordination of EO information to forecast and prepare for environmental hazard events, thereby reducing damage and increasing disaster resilience and recovery capabilities.

Committee on Earth Observing Satellites (CEOS)

CEOS was established in September 1984 in response to a recommendation from a panel of experts on remote sensing from space, set up under the G7 Economic Summit of Industrial Nations Working Group on Growth, Technology, and Employment. This panel emphasizes the value of coordinating international EO efforts through CEOS to benefit society.

98 NCEI. Maps and Geospatial Products.

99 Group on Earth Observations (GEO).

100 GEO Knowledge Hub.

CEOS currently consists of 34 member agencies, comprising the main space agencies of the world, together with 30 associate member agencies covering a range of geospatial and geoscience capabilities.[101] CEOS focuses on international coordination of civilian space-based EO programs and development activities. CEOS promotes sharing of EO data to increase societal benefit and better inform decision-making for a sustainable future.

CEOS working groups address topics such as calibration and validation, data portals, capacity building, disaster management, climate, and common data processing standards shared across a wide range of EO domains.

CEOS virtual constellations coordinate space-based, ground-based, and/or data delivery systems to meet a common set of requirements within seven specific domains.

CEOS Committee on Earth Observation Satellites

▲ The CEOS 2023 EO Handbook (image from the Committee on Earth Observing Satellites).

The *CEOS Earth Observation Handbook*, 2023 edition, coincides with the first Global Stocktake (GST) of the Paris Agreement.[102] GST is a data-driven process with the goal to more effectively formulate actions to mitigate and to adapt to climate variability. GST aims to use the information gathered on the state of key planetary indicators and on the impact of human efforts to reduce greenhouse gas (GHG) emissions. Satellite EO is fundamental to the measurement and production of objective facts and figures that are required. In this context, the CEOS EO Handbook is a pragmatic guide for all stakeholders engaged in the climate policy process to help them understand how satellite EO data can assist them.

United Nations Committee of Experts on Global Geospatial Information Management (UN-GGIM)

UN-GGIM is promoting novel approaches to geospatial data management by implementing a global policy framework that will help countries to better integrate geospatial and other key information into both global development policies and into their own national development plans.[103]

Established in 2011, UN-GGIM defines guidelines for the production and use of geospatial information within national and global policy frameworks, and for the capacity building of geospatial information technical skills, especially of developing countries.

UN-GGIM

101 Committee on Earth Observation Satellites (CEOS). Agencies.

102 The Earth Observation Handbook.

103 UN. UN Global Geospatial Information Management (UN-GGIM).

UN-GGIM partners with GEO, CEOS, and other international organizations to build a global, interconnected data ecosystem that will allow for more coherence and consistency in planning and implementation of actions to achieve Sustainable Development Goals (SDGs).

The *UN Sustainable Development Goals Report* (2023 Special Edition) provides an overview of progress on the implementation of the 2030 Agenda for Sustainable Development, using the latest available data and estimates.[104] It tracks both global and regional advances toward all of the 17 goals, with in-depth analyses of selected indicators for each goal.

2.6 Current Overall Earth Observation Market Estimates and Growth Forecasts

There are various estimates available for the current and future size and composition of the EO data and analytics market.

This section summarizes the key findings from two of the most recent and comprehensive EO market reports prepared by EuroConsult (*Earth Observation Data & Services Market,* 16th edition, published in December 2023) and the European Union Space Programs Agency *(EUSPA EO and GNSS Market Report Issue 2,* published in January 2024).

2.6.1 EuroConsult Market Estimates

As global economies continue to navigate the challenges posed by inflation, geopolitical tensions, and the residual effects of pandemic-related disruptions, the EO market is now undergoing a highly dynamic and transformative phase, presenting both opportunities and obstacles for industry players worldwide. Commercial operators are restructuring, reducing costs, and optimizing resources with a greater customer focus on execution. Security, sustainability, and digitalization are key factors driving market dynamics.[105]

The EO data market currently stands at $1.78 billion (as of 2022), growing at a 5-year compound annual growth rate (CAGR) of 6%. Forecasts reveal future growth to reach a total value of more than $2.7 billion by 2032, growing at a 4% CAGR throughout the decade, mostly driven by the demand for very-high-resolution (submetric) imagery.

In parallel, the EO services and analytics market is currently valued at $2.86 billion (as of 2022) and demonstrates a 5-year CAGR of 9%. This market is predicted to increase in value, totaling $4.9 billion by 2032 and growing at a 6% CAGR. The commercial value-added services sector, given its highly diverse products and services, remains fragmented, with the top two players contributing approximately 26% of revenues (28% in 2021). The business-to-government sector continues to be the core element, with defense as the most significant sector, valued at $2 billion in 2022.

[104] UN. UN Statistics Division (UNSD).

[105] Novaspace. 2023. Global market for commercial Earth Observation data and services to reach $7.6 billion by 2032. 28 November.

Figure 14: Earth Observation Data and Services Market, 2022

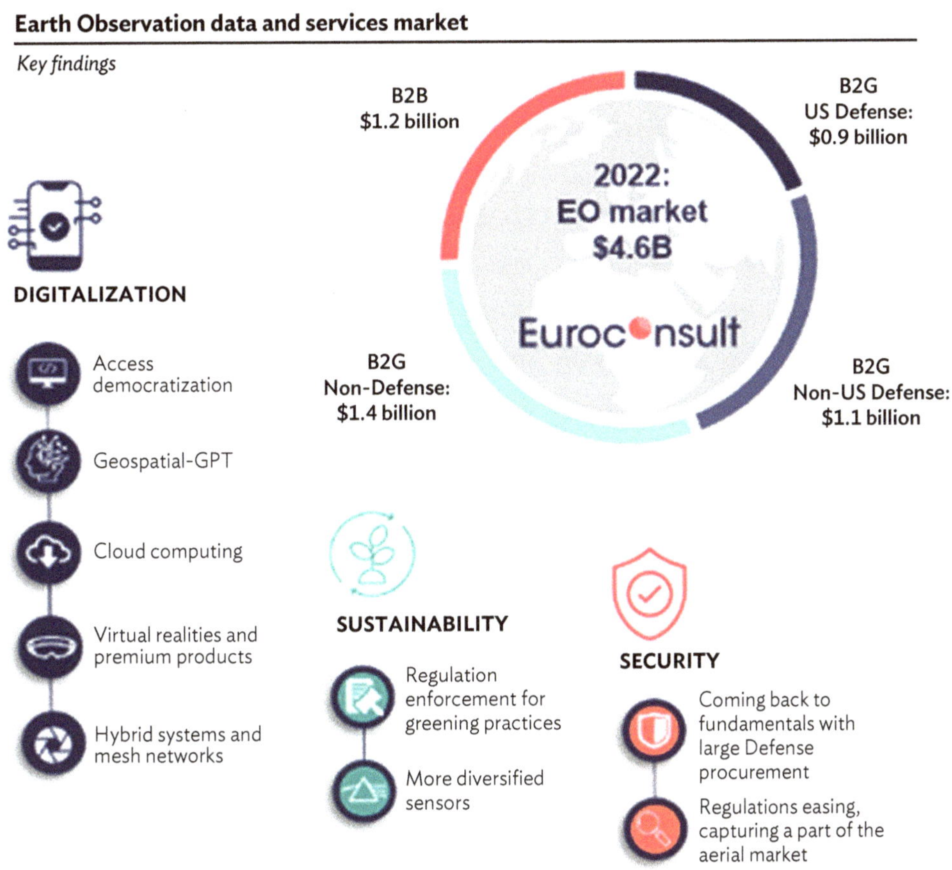

Source: Novaspace. 2023. Global market for commercial Earth Observation data and services to reach $7.6 billion by 2032. Press release. 28 November.

In terms of geographic distribution, North America retains its position as the market leader in EO, comprising nearly 45% of total global revenue. Europe, accounting for 22% of the market share, witnessed a shift in demand from data to informational services based on more cost-efficient supplies. Asian markets, led by the PRC, Japan, and the Republic of Korea, are now ready for a period of strong growth, driven by domestic use and expanding export potential.

2.6.2 EUSPA Market Estimates

The EO market is expected to continue to grow rapidly from €3.4 billion in 2023 to almost €6 billion in 2033, including both data and value-added service revenue. This growth is driven by increasing attention to environmental sustainability and the reduction of the negative impact of human activity; the improved, more customer-focused EO service offerings; and increasing customer awareness on the potential use and added value of EO information.

Revenues generated from EO data alone in 2023 amounted to almost €600 million across all segments. The EO data market is expected to see a CAGR of 5% by 2033, resulting in total revenues of almost €1 billion.

The EO value-added services market generated a global total of around €2.8 billion in 2023. It is expected to achieve a CAGR of almost 6%, growing total revenues to almost €5 billion by 2033.

Almost half of this total revenue is generated by the top three market sectors: climate environment and biodiversity, agriculture and urban development, and cultural heritage. However, projections suggest significant future growth in the insurance and finance segment, with an expected increase from around €340 million in 2023 to almost €900 million in global EO revenues in 2033, making it the leading revenue generator.

Figure 15: Distribution of Earth Observation Market Revenue by Market Segment, 2023

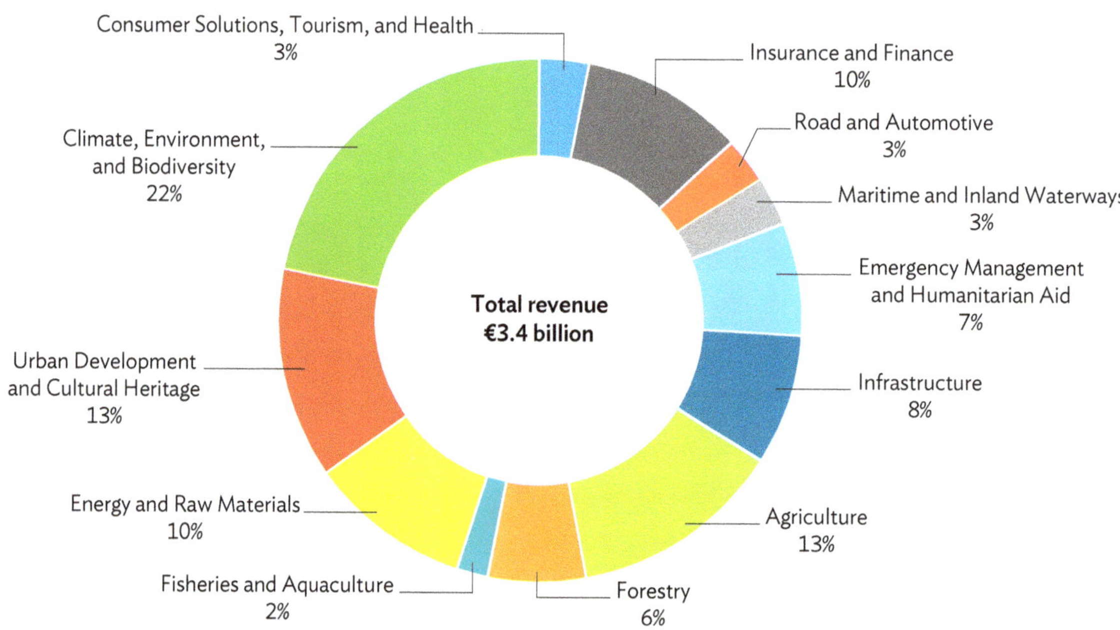

Source: European Space Programmes Agency (EUSPA). 2024. *EO and GNSS Market Report*. 2024 edition, p. 12.

In terms of geographic distribution, Europe and the US generate the bulk of revenues from both EO data and value-added services.

Collectively, European and US companies account for a market share of more than 85%, with each being responsible for more than 40%. Companies from the PRC represent 6% of the market, while Canada and Japan generate 3% and 2% of global revenues, respectively.

In summary, both EuroConsult and EUSPA estimates for the EO services and analytics market largely concur both in terms of current and forecasted growth, while for the EO data market, there are differences, with the EuroConsult estimates being significantly higher.

North America is the largest geographic region in terms of EO market share. The region's enabling factors for the uptake of EO include a high number of ongoing national research studies and capital investments in the EO sector, the implementation of advanced space programs and development of digital infrastructure, and the high adoption rate of commercial satellite imaging across end-user industries nationally. However, Asia and the Pacific is estimated to grow at the highest CAGR over the period of 2023–2027.[106]

Figure 16: Geographic Distribution of Earth Observation Market Revenue, 2023

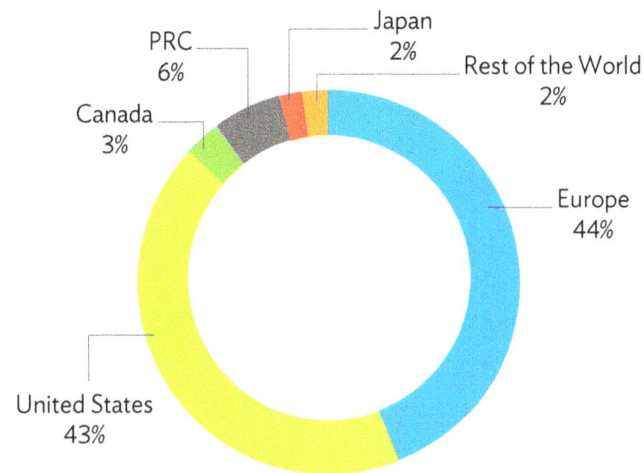

PRC = People's Republic of China.

Source: European Space Programmes Agency (EUSPA).

The next big growth areas in the EO market are forecast to be in the following two separate sectors, but also connected through the common themes of climate variability and extreme events:

(1) Integration of EO data and information into private sector global corporate companies driving the digital transformation of their climate variability proofing.[107]

(2) Insurance and finance market segment serving insurers (and re-insurers), and international and local financial institutions (e.g., private and commercial banks, investors and stock traders).

There are a number of innovative financing models related to the broader area of sustainable development that represent opportunity for further growth in the EO market. Examples include the following:

(1) Parametric insurance products for local-level solutions where insurance payouts are triggered by a pre-established set of thresholds for defined environmental parameters such as a storm reaching a particular windspeed and/or rainfall volume. Through parametric insurance, the payouts are not related directly to the loss incurred, thereby accelerating the time between the damage event occurring and the reconstruction.

(2) Risk pooling or risk sharing for regional-level solutions wherein insurance players agree to distribute the financial damages caused by environmental hazards for large areas where the overall probability and occurrence of hazard events is well known, but the exact location is more difficult to predict. The Caribbean Catastrophe Risk Insurance Facility is the world's first multi-country regional risk pool. It offers access to quick liquidity following disasters in the region.

(3) Green bonds which are similar to regular bonds (debt financing) but are raised from investors to finance a range of development projects that have a positive impact on the environment (e.g., renewable energy, reducing carbon emissions, smart cities). The UN SDGs have helped accelerate the expansion of the green bond market significantly over the last decade.

[106] Mordor Intelligence. *Satellite-Based Earth Observation Market Size and Share Analysis: Growth Trends and Forecasts (2025–2030).*

[107] Sifted. *Spacetech.*

Annual issuance of all green, social, sustainability, and sustainability-linked bonds could hit $1.05 trillion in 2024, up from $0.98 trillion in 2023, according to S&P Global.[108]

(4) Financing for nature-based solutions for the conservation and restoration of environmental ecosystems where the UN Environment Programme calls for the tripling of financing for nature-based solutions to $542 billion by 2030. This includes innovative financing schemes for carbon-market approaches (i.e., transfer or trade-in unitized mitigation outcomes [carbon credits]) and non-carbon-market approaches (i.e., for project sites where carbon crediting is not a viable or preferable financing option).

(5) Payment for Ecosystem Services (PES) where financial incentives are offered to actors to encourage them to not cause environmental harm (e.g., to farmers or landowners who have agreed to take specifically designed actions to manage their land or watersheds to provide an ecological service).[109] Coastal and marine resources provide the world with a range of critical ecosystem services, yet globally, these resources are fast diminishing. A promising financing strategy to reduce this decline in coastal resources could be to establish PES schemes or to incorporate an element of PES into existing regulatory mechanisms.

2.7 Benefits of Using Satellite Earth Observation Information

EO satellite technology serves as a source of key thematic types of environmental information that is global, comprehensive, accurate, repeatable, and timely. More importantly, within the context of regional development assistance, these same EO data can increase the capacity of developing countries and regions to acquire, analyze, and utilize a more comprehensive and up-to-date environmental information, thereby growing their digital economies in the private sector and enabling better policymaking in the public sector.

2.7.1 Top-Down Benefits Estimation Methods

There are a number of top-down analyses to estimate the economic, social, environmental, and strategic benefits for satellite EO in general and specific missions to the public and private sectors, as well as to the wider community of final users.

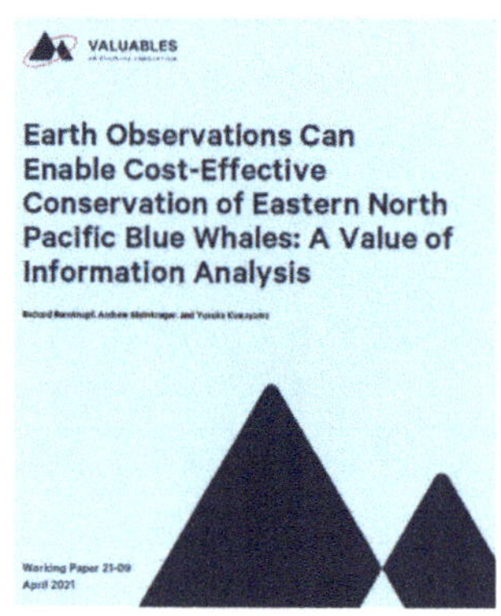

Resources for the Future, an independent, nonprofit research institution based in Washington DC, together with NASA, is measuring the socioeconomic benefits that satellite EO provides to manage water resources, health and air quality, climate variability, wildfires, and more. The benefits are estimated via a series of impact assessments. These are quantitative studies that use a thorough methodology to explore how people use improved environmental information to make better decisions. The studies quantify how these decisions then impact defined socioeconomic metrics such as lives saved or resources conserved.[110]

---

[108] *World Economic Forum.* 2024. *What are green bonds and why is this market growing so fast?* 22 November.

[109] International Institute for Environment and Development. *Markets and payments for environmental services.*

[110] Resources for the Future. Impact Assessments.

The European Commission has published several fact sheets that define and demonstrate quantifiable, practical examples of benefits achieved by using the Copernicus Sentinels data. The types of benefits looked at comprise cost reductions, additional revenues, or environmental improvements. Each fact sheet addresses a specific sector including agriculture, forestry, insurance, ocean monitoring, oil and gas, renewable energies, urban monitoring, and air quality.[111]

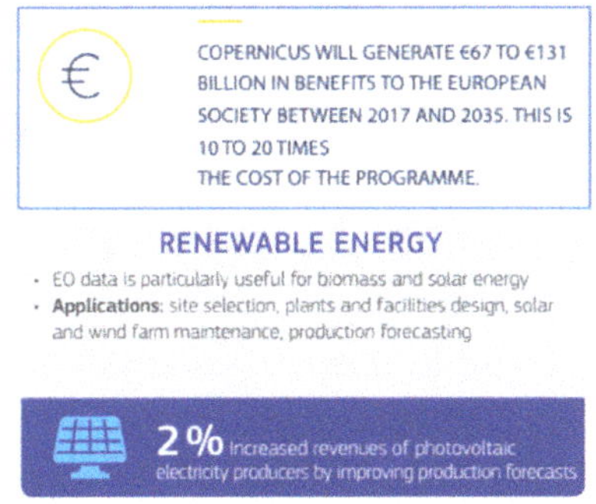

2.7.2 Bottom-Up Benefits Estimation Methods

A good example of a systematic approach and definition of a detailed bottom-up methodology in estimating the benefits of using satellite EO information is that of the Sentinels Benefits Study, carried out by the European Association of Remote Sensing Companies (EARSC) and its partners between 2017 and 2021, and commissioned by the European Space Agency (ESA).[112]

The core of the Sentinels Benefits Study is to understand how a specific EO-derived service is being used by an organization, and how the integration of EO into activities of the organization are benefiting third parties that they deal with, and ultimately society and citizens at large. The approach is bottom-up based on a detailed evaluation of a value chain.

Figure 17: Four Tiers of Earth Observation Value Chain Identified for Benefits Components Analysis

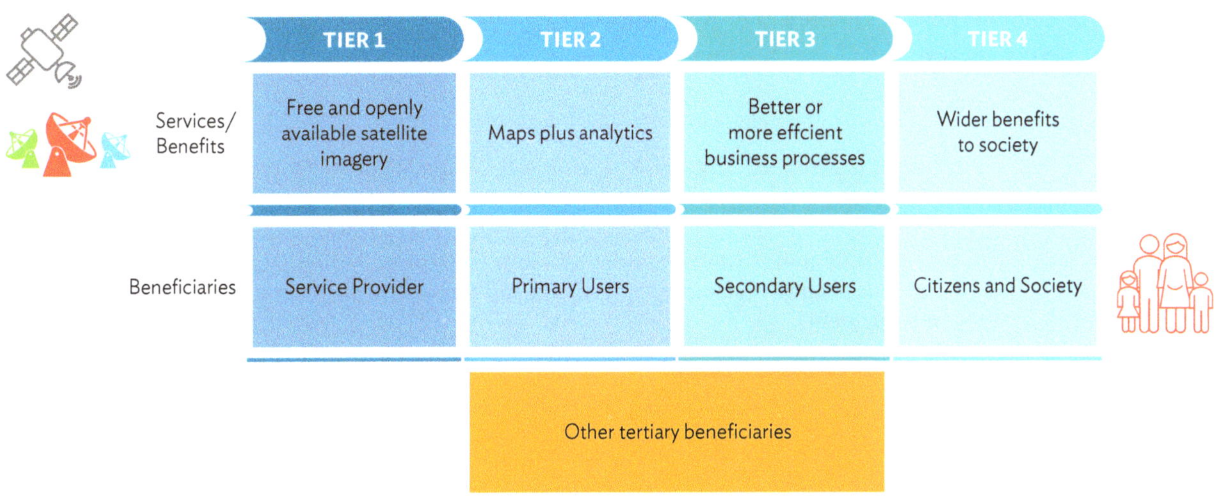

Sources: European Association of Remote Sensing Companies, European Space Agency, and European Union. 2024. *SeBS Methodology: A Practical Guide for Practioners to evaluating the benefits derived from the use of Earth Observation data*. p. 22.

111 Copernicus. Benefits.

112 EARSC. Sentinel Benefits Study (SeBS).

To focus and structure the analysis, a set of six dimensions of EO benefits are defined, which may or may not be monetized (i.e., each dimension may be assessed quantitatively in financial terms or qualitatively or quantitively in nonfinancial terms as shown in Figure 18).

Figure 18: Six Dimensions of Earth Observation Benefits (Monetized and Non-monetized)

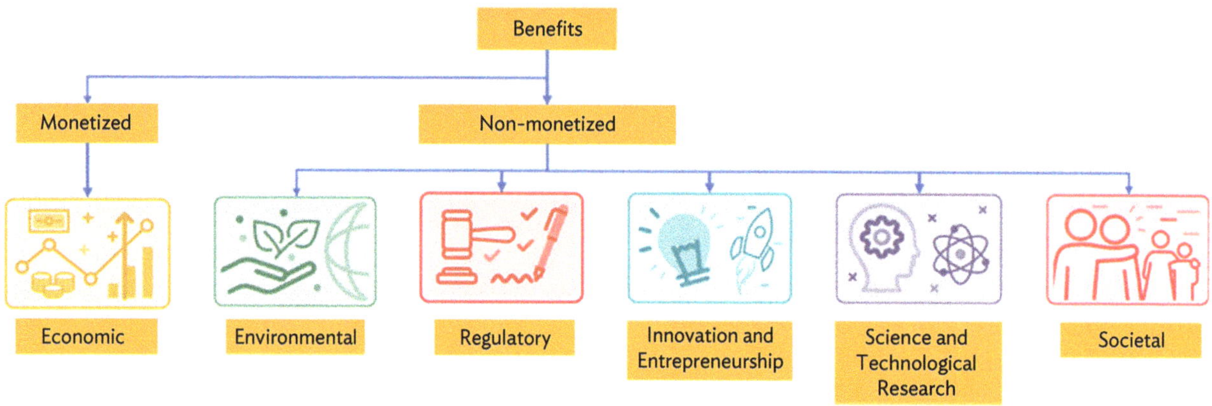

Source: European Association of Remote Sensing Companies, European Space Agency, and European Union. 2024. *SeBS Methodology: A Practical Guide for Practioners to evaluating the benefits derived from the use of Earth Observation data*. p. 25.

To guide the benefits evaluation and to improve consistency between analyses, a set of indicators has been introduced for each of these six dimensions along with a quantitative scale of five specific levels on which their impact can be represented.

Table 4: An Example of Indicators for a Monetized Earth Observation Benefit (the Economic Domain)

Economic	**Impacts related to the production of goods or services, or impacts on monetary flow or volume, such as revenue, profit, capital and (indirectly, through turnover generation) employment.**
Capital expenditure reduction or avoidance	Aversion/reduction of capital expenditure which would or could otherwise have been incurred
Cost savings	Aversion/reduction of operating expenditure which would or could otherwise have been incurred
Increased revenues	Income through the commercial exploitation of EO-enabled value-added services
Efficiency gains	Time or material saved through the use of services
Reduction of risk	Reduction of risk and consequential costs
Employment	Increased employment in the service provider (which is not a cost in the value chain)

EO = Earth Observation.

Source: European Association of Remote Sensing Companies, European Space Agency, and European Union. 2024. *SeBS Methodology: A Practical Guide for Practioners to evaluating the benefits derived from the use of Earth Observation data*. p. 30.

Table 5: An Example of Indicators for a Non-monetized Earth Observation Benefit (the Regulatory Domain)

Regulatory	**Impacts linked to the development, enactment or enforcement of regulations, directives and other legal instruments by policymakers.**
Better regulations	Improvements in the policymaking process and the resultant policy as a result of knowing that Earth Observation data is available to support
Improved monitoring	An improved ability to monitoring compliance or implementation with respect to regulations (e.g., more efficient monitoring or CAP compliance)

CAP = Common Agricultural Policy of the EU.

Source: European Association of Remote Sensing Companies, European Space Agency, and European Union. 2024. *SeBS Methodology: A Practical Guide for Practioners to evaluating the benefits derived from the use of Earth Observation data*. p. 38.

This Sentinels Benefits Study methodology has been applied to 18 in-depth case studies ("Long Reports") and 10 first analysis studies ("Short Reports"), each addressing different EO thematic and user sectors.[113]

▲ Example of case studies in the Sentinels Benefits Study (images from European Association of Remote Sensing Companies, European Space Agency, and European Union).

One example of these case studies is on sustainable palm oil production using Sentinel-1 and Sentinel-2 data to constantly monitor forested areas in Southeast Asia.[114] The service creates "deforestation alerts," which notify clients in the wood trade that deforestation is occurring on forest plantations that they are using as a source for buying wood products.

[113] EARSC. Case Gallery.

[114] EARSC. *Deforestation Monitoring for Sustainable Palm Oil Production.*

The Roundtable on Sustainable Palm Oil has established the certification standards for sustainable palm oil and fully endorses and encourages the use of EO data as part of the certification procedure. One of the many economic benefits of implementing sustainable palm oil production is that food companies can gain access to sustainability-linked loans. Through these loans, the interest rates can be directly tied to a company's sustainability performance and activities, with EO being used for monitoring and evaluation.

Based on the Sentinels Benefits Study, the total volume of benefits in this sector was estimated to be €19 million to €33.6 million per year.

Figure 19: Total Estimated Volume of Earth Observation Benefits for Sustainable Palm Oil Production in Southeast Asia via the SeBS Methodology

pa = per annum, SeBS = Sentinels Benefits Study.

Source: Image from European Association of Remote Sensing Companies, European Space Agency, and European Union. Deforestation Monitoring for Sustainable Palm Oil Production.

An example of a non-monetized benefit in the regulatory dimension is that of the European directive 2007/2/EC, Infrastructure for Spatial Information in Europe (INSPIRE directive).[115]

The INSPIRE Directive aims to create an EU Spatial Data Infrastructure (SDI) to support and implement EU environmental policies and activities that have an environmental impact. This European SDI will simplify public access to spatial information and help in policymaking across thematic sectors and across Europe via the sharing of environmental spatial information among public sector organizations. Hence, it is especially relevant for those countries that are at the initial stages of planning or developing a national spatial data infrastructure.

The INSPIRE Directive is a key legal instrument ensuring that the necessary data is available to address environmental and climate challenges (i.e., A Green Deal for Europe [green transition]) and strengthen the emerging data economy (i.e., European Common Data Spaces [digital transition]).

The INSPIRE Conference, *Green Data for All*, held on 28–29 November 2023, focused on bridging the gap between the green and digital sectors by bringing together various thematic communities.[116]

[115] Infrastructure for Spatial Information in Europe (INSPIRE).

[116] INSPIRE. INSPIRE Conference 2023.

Of relevance to the development context are the INSPIRE Smart Cities Marketplace, the 2030 Biodiversity Strategy, the Zero Pollution Action Plan, and the Climate Change Adaptation Strategy, which are all part of the INSPIRE Green Deal Data Space.[117]

Although INSPIRE is a non-monetized benefit dimension, the specific benefits to citizens, businesses, and government agencies have been explored in detail for many of the 34 data themes that this European EU directive encompasses.[118]

2.8 Future Trends in the Earth Observation and Geospatial Information Landscape

A comprehensive analysis of the *Main Trends and Challenges in the Space Sector* has recently been produced by PwC.[119]

The publication provides a complete and detailed definition of space sector trends and challenges from macro to domain-specific levels. The main space domains addressed include EO, access to space, satellite communication, global navigation, space security and situational awareness, and in-orbit economy and space exploration.

Specifically, the EO sector is now experiencing far-reaching changes that will have strong impacts on the EO industry, as the EO market is rapidly evolving and attracting a growing number of external investors, notably in the US.

The main demand and supply trends are identified in Table 6.

117 Smart Cities Marketplace; INSPIRE. The Great Project.

118 wetransform. The Benefits of INSPIRE.

119 PwC. 2022. *Main Trends and Challenges in the Space Sector.* Third Edition.

Table 6: Main Earth Observation Demand and Supply Trends, 2022–2023

Demand Trends	Supply Trends
Democratization of the use of EO outside of expert communities	Toward vertical integration and satellite miniaturization
Growing interest in vertically specialized products and insights	Constellations are the new norm for most players
Strong interest in advanced analytics from military	Expected consolidation of the EO industry, notably due to SPACs
New emerging analytics markets	Emergence of vertical champions on key domains
Growing interest for non-imagery EO data and information	Emergence of major non-imagery players

EO = Earth Observation, SPAC = special purpose acquisition company.

Source: PwC. 2022. *Main Trends and Challenges in Earth Observation*. 3rd edition. p. 41.

Some of the key highlights in terms of opportunities for growth within the EO sector in the short term include the following:

- Major players in the insurance sector are now investing significantly in satellite EO analytics to obtain specific environmental information tuned to their business operations (e.g., AXA Climate, SwissRe, MunichRe).
- There is a growing and strong interest from the agri-food industry to improve transparency of agri-production processes and food traceability, particularly in the use of chemical fertilizer.
- There is a growing interest in environmental sustainability and GHG emission monitoring, which is expected to stimulate the demand for EO-based information solutions.
- There is an increasing direct intervention of investment banks and capital into the space sector, encouraging companies in the EO business to announce very ambitious growth plans.

In 2022, GeoSpatial World identified the top five trends in EO.[120] The emergence of as-a-service models is a key recent development that will continue and strengthen in the future.

In particular, the space-as-a-service models allow governments to participate in the development of EO missions by supplying the critical payload, while essentially outsourcing the rest of the space segment to commercial companies. These include companies such as Spire and Loft Orbital, which are responsible for payload integration, assembly and testing, launch service agreements, and satellite operation.

[120] A. Ravichandran. 2022. Top Five Trends in Earth Observation. *Geospatial World*. 7 September.

This model enables the private sector to contribute to the design of the mission, as well as to the manufacturing of payload, ensuring that the economic rationale for the EO mission is guaranteed.

Satellite-as-a-service models are a recent development which allows countries and companies to acquire ready-made satellites from manufacturers, ensuring that the assets are fully owned by customers. One example is that of ICEYE, a Finnish microsatellite SAR player, which announced recently that it would deliver its new offering of fully operational SAR satellites in the coming months.

Completing the stack of as-a-service models are the data-as-a-service, platform-as-a-service, and insight-as-a-service, where rapid evolution is taking place with go-to-market strategies being employed by commercial EO companies as they scale up and focus on the impact of EO on the customers and how it helps them get their job done.

Figure 20: Earth Observation 'Stack' with 'As-a-Service' Business Models Identified

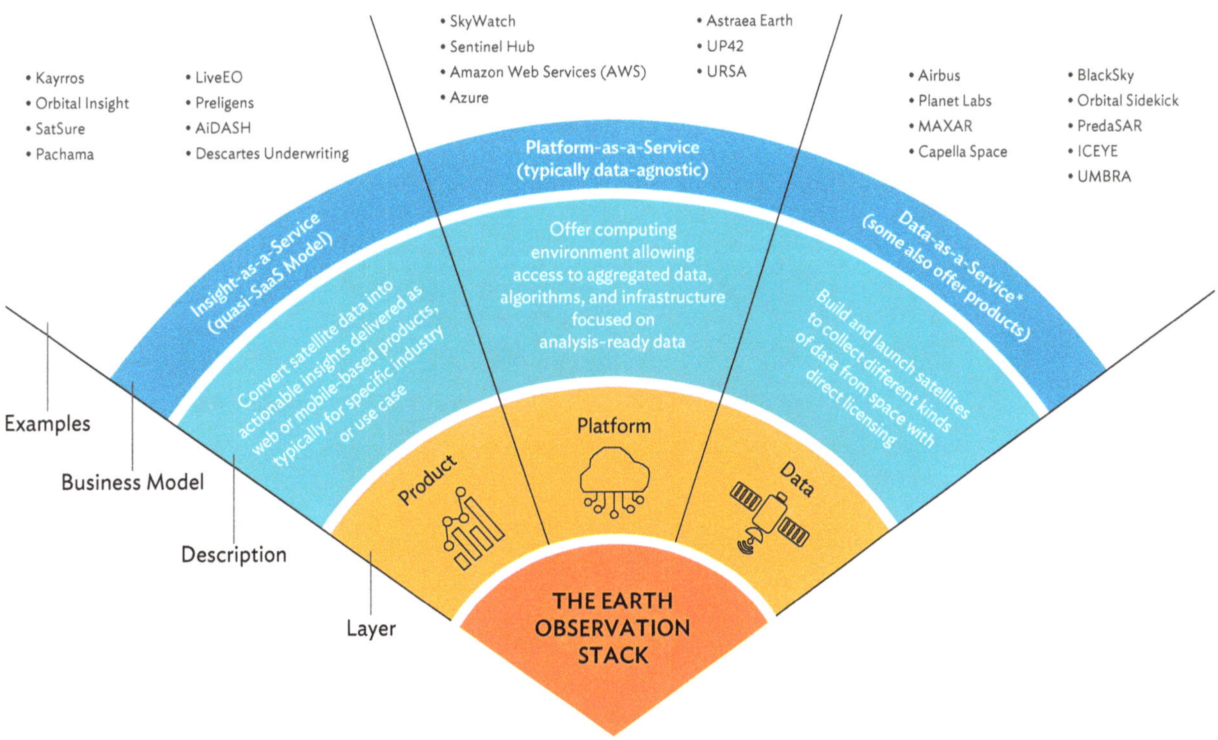

Source: Aravind. 2022. The State of Commercial Earth Observation: 2022 Edition. *TerraWatch Space.* 28 August.

In broader geospatial information context, the United Nations Committee of Experts on Global Geospatial Information Management (UN-GGIM) published a report in August 2020 on *Future Trends in Geospatial Information Management: The Five-to-Ten-Year Vision.*[121]

[121] C. Walter. 2020. *Future trends in geospatial information management: The five to ten year vision.* Third Edition.

The report presents an in-depth analysis of key topics, including drivers and trends in geospatial information management; the digital infrastructure of the future; the rise of AI, big data, and big data analytics; and trends in technology and the future direction of data creation, and maintenance and management.

The report identifies the top trends that are likely to affect the geospatial industry over the upcoming decade. The specific trends have been assigned to be associated with one of five top-level industry drivers. Through analyzing these trends, forecasts of how these top-level drivers are expected to evolve over the next 5 to 10 years can be derived. The impacts of these trends have been characterized by estimating the level of predictability and their potential impact on the geospatial industry (Figure 21).

Overall, this high-level analysis shows that technological developments and the availability of new data sources and analytical methods are important, but the geospatial information industry is influenced by a much wider set of drivers. A data-driven society will spark innovation which is central to growth. This innovation will be further reinforced by improvements in the areas of connectivity, sensor networks, data analytics, and cloud computing.

Figure 21: Chart of the Predictability and Impact of Future Prevailing Drivers and Underlying Trends in the Geospatial Information Industry

Five drivers will advance change in the global geospatial information management landscape over the next 5 to 10 years

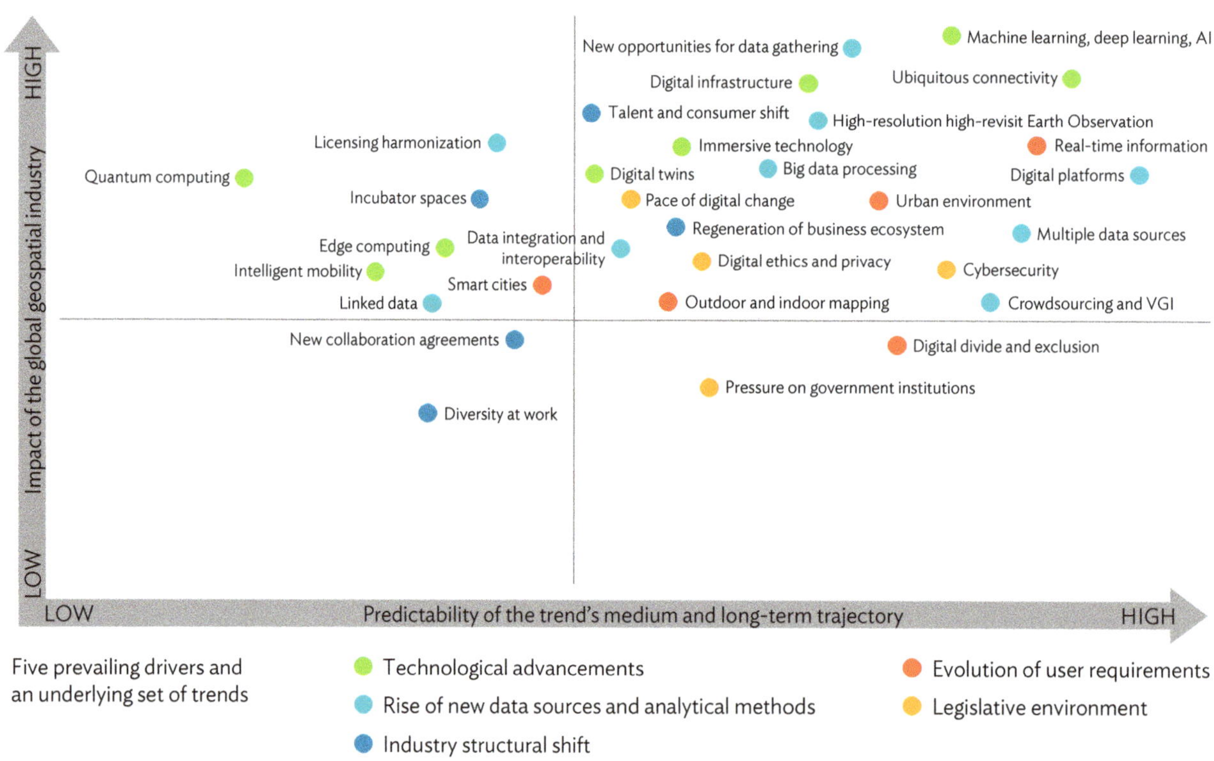

Source: United Nations Committee of Experts on Global Geospatial Information Management (UN-GGIM). 2020. *Future Trends in geospatial information management: the five to ten year vision*. Third edition. p. 17.

In summary, as the global economy continues to navigate the challenges posed by inflation, geopolitical tensions, and the residual effects of pandemic-related disruptions, the EO sector is undergoing a rapidly transformative phase, with commercial operators restructuring, reducing costs, and optimizing resources, with a strong focus on execution in an increasingly competitive landscape.

Across the broader geospatial industry, the next 5 to 10 years will see significant developments in the maturity and application of already well-established technologies. Among others, AI, sensor technology, and the Internet of Things will drastically change the way in which data are collected, managed, and maintained.

Economies of all sizes will benefit from more affordable drone and satellite technologies and image analytics capabilities as data sources become more numerous, accessible, and readily used. This increased use of digital data technology and analytical methods will reduce the geospatial digital divide between economies of different income levels over the decade to come.[122]

122 G. Scott. 2019. *The Integrated Geospatial Information Framework Bridging the Geospatial Divide.* Presentation prepared for the First International Workshop on Operationalizing the Integrated Geospatial Information Framework. 9–11 September.

USE CASES OF SATELLITE EARTH OBSERVATION IN ADB DEVELOPMENT OPERATIONS

3

There are now many examples of the uses of satellite EO environmental information in supporting development, particularly through the activities of the European Space Agency (ESA) in cooperation with the Asian Development Bank (ADB), European Investment Bank, Inter-American Development Bank, International Fund for Agricultural Development, Global Environment Facility, and the World Bank.

Twelve targeted demonstrations explored the use of EO-based environmental information in a set of operational development projects being implemented by ADB. The results can be found in the report, *Earth Observation for a Transforming Asia and Pacific*, published in 2017 by ESA in partnership with ADB.[123]

Earth Observation Use Cases in the Context of ADB Operations

In this section, six specific and recent EO use cases are presented in support of ADB development operations, covering the following thematic areas:

(1) Basic land surface type (cover) mapping and monitoring.

(2) Disaster risk management and disaster risk reduction.

(3) Climate resilience and proofing.

(4) Coastal zone management.

(5) Forest monitoring and management.

(6) Greenhouse gases monitoring (CO_2, CH_4, N_2O).

A more detailed description of the types of EO information content that can be delivered in these six thematic areas is given below, together with a summary description of how this EO information is relevant and can be used in the development context.

This section concludes with an additional EO use case that is focused not on ADB development operations, but on how EO-based information can help monitor and understand changes in the urban environment related to people's access to basic services and adequate, safe, and affordable housing (SDG Target 11.1).

[123] ADB and ESA. 2017. *Earth Observation for a Transforming Asia and Pacific.*

3.1 Earth Observation Use Case 1: Basic Land Surface Type (Cover) Mapping and Monitoring

The main EO information content consists of basic mapping services to support additional analytics (e.g., disaster risk management, agriculture, urban).

- Land surface type (cover) mapping typically provides information about basic forest types, major agricultural surface types, conservation areas, settlements, infrastructure, primary roads, bare soil, water bodies, rivers, and wetlands based on a standard classification scheme (e.g., in accordance with the European Commission Coordination of Information on the Environment [CORINE] Land Cover [CLC – 44 thematic classes], or the UN Food and Agriculture Organization Land Cover Classification System [LCCS – 8 major land cover types are defined]). For Indonesia, 13 specific categories of surface type were used (e.g., settlements, cultivation, paddy fields, tree cover, tree crops, mangrove, shrub, grassland, sparse vegetation, savanna, bare soil, wetland, and water bodies).

Map 2: Land Cover Map of Cianjur, West Java, 2020–2021

Note: This map was captured through processing of multi-sensor satellite images.

Source: Planetek Italia, reference: ADB. 2023. *Earth Observation Services and Tools for Development: Examples from Indonesia*. p. 14.

Map 3: Land Use and Land Cover Maps of Buutsagaan, Zag, and Khureemaral Soums in Central Mongolia, Province of Bayankhongor; September 2013 and September 2014

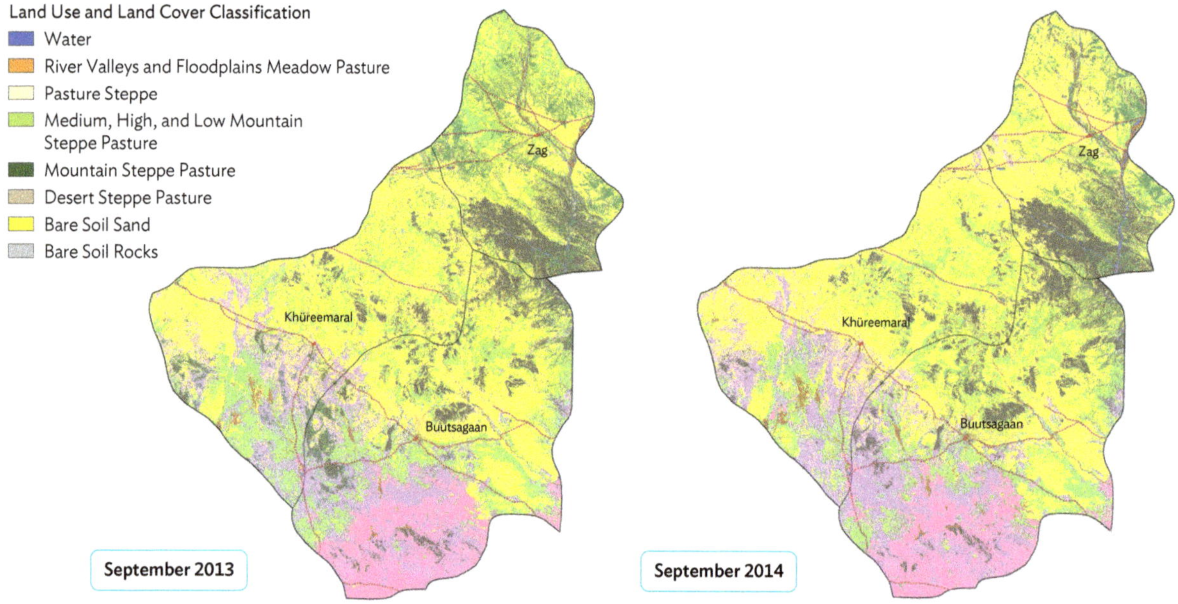

Note: These maps were processed from LandSat-7 and LandSat-8 multispectral image data at 30 m resolution.

Sources: DEIMOS Engenharia SA, Portugal; European Space Agency, reference: ADB. 2017. Earth Observation for a Transforming Asia and Pacific. p. 13.

Map 4: Land Use and Land Cover Changes in Buutsagaan, Zag, and Khureemaral Soums in Central Mongolia, Province of Bayankhongor, 2013–2014

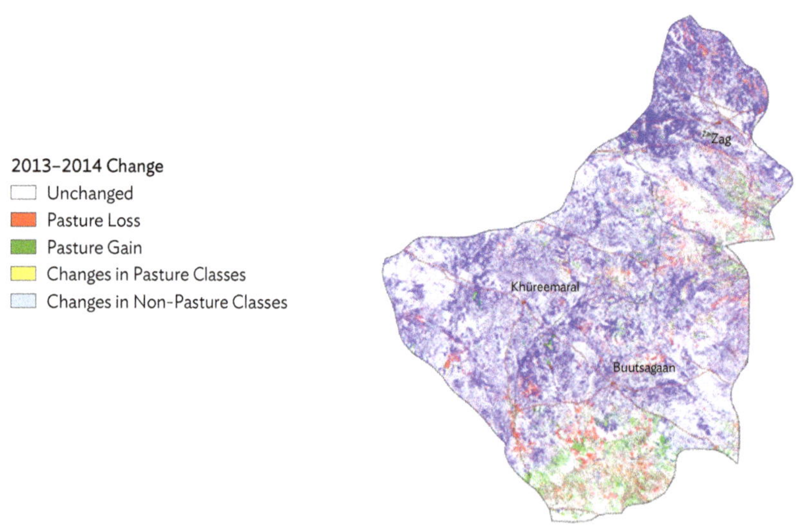

Note: These maps were processed from LandSat-7 and LandSat-8 multispectral image data at 30 m resolution.

Sources: DEIMOS Engenharia SA, Portugal; European Space Agency, reference: ADB. 2017. Earth Observation for a Transforming Asia and Pacific. p. 13.

- Vegetation and biomass indexes mapping. These indexes are for monitoring the state and evolution of biomass health, vigor, and productivity, and can serve as early warning indicators (e.g., Normalized Difference Vegetation Index, Anthocyanin Reflectance Index, Enhanced Vegetation Index, Normalized Difference Moisture Index, Normalized Difference Water Index, Soil-Adjusted Vegetation Index, Leaf Area Index, Modified Anthocyanin Reflectance Index, Green Normalized Difference Vegetation Index, Modified Chlorophyll Absorption Ratio Index, Normalized Difference Chlorophyll Index, and Fraction of Absorbed Photosynthetically Active Radiation).

Map 5: Normalized Difference Vegetation Index (top) and Soil-Adjusted Vegetation Index (bottom) for the Province of Nusa Tenggara Timur

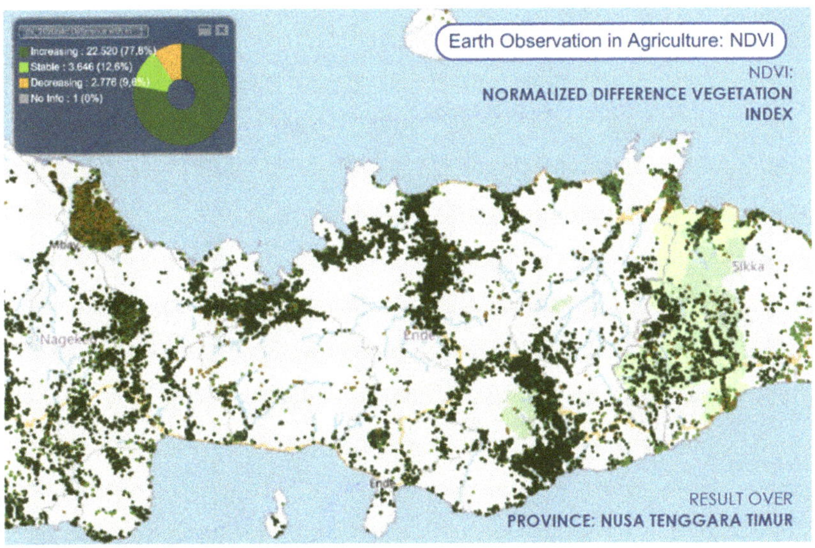

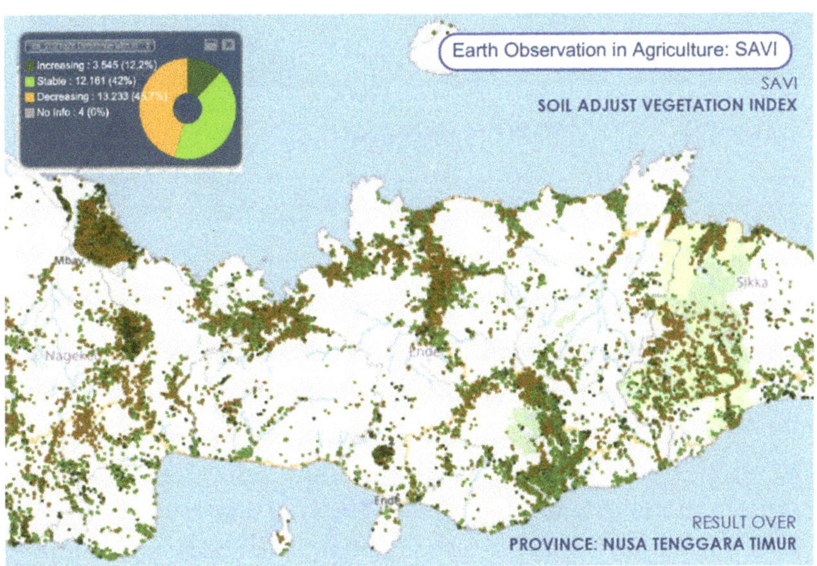

Source: Planetek Italia, reference: ADB. 2023. *Earth Observation Services and Tools for Development: Examples from Indonesia*. p. 13.

- Reference mapping using satellite archived EO data or recent acquisitions. Minimum information content of reference mapping products includes the distribution of roads, rivers, and highly populated urban and peri-urban areas, including land surface height information (topography) and administrative boundaries.

Map 6: Reference Mapping of Banda Aceh, Sumatra, After December 2004 Tsunami

Source: European Space Agency, reference: European Space Agency. 2013. *Earth Observation for Green Growth.* p. 42.

This city map was annotated with in-situ information which located the main recovery infrastructure, medical centers, and other humanitarian structures. It provides the standard required base-layer information as a standardized geographical reference dataset that can be used to determine the key geographical attributes of a given area.[124]

- Up-to-date asset mapping of urban areas, including government assets (schools, hospitals, roads), large or critical infrastructures (transport, pipelines, ports), residential and commercial and industrial exposure (includes building footprints and building heights), and changes in urbanization patterns (growth and expansion, slum growth, vertical changes).

[124] ESA. 2013. *Earth Observation for Green Growth.* p. 42.

Map 7: Urban Land Use Information in Mandalay City, Myanmar

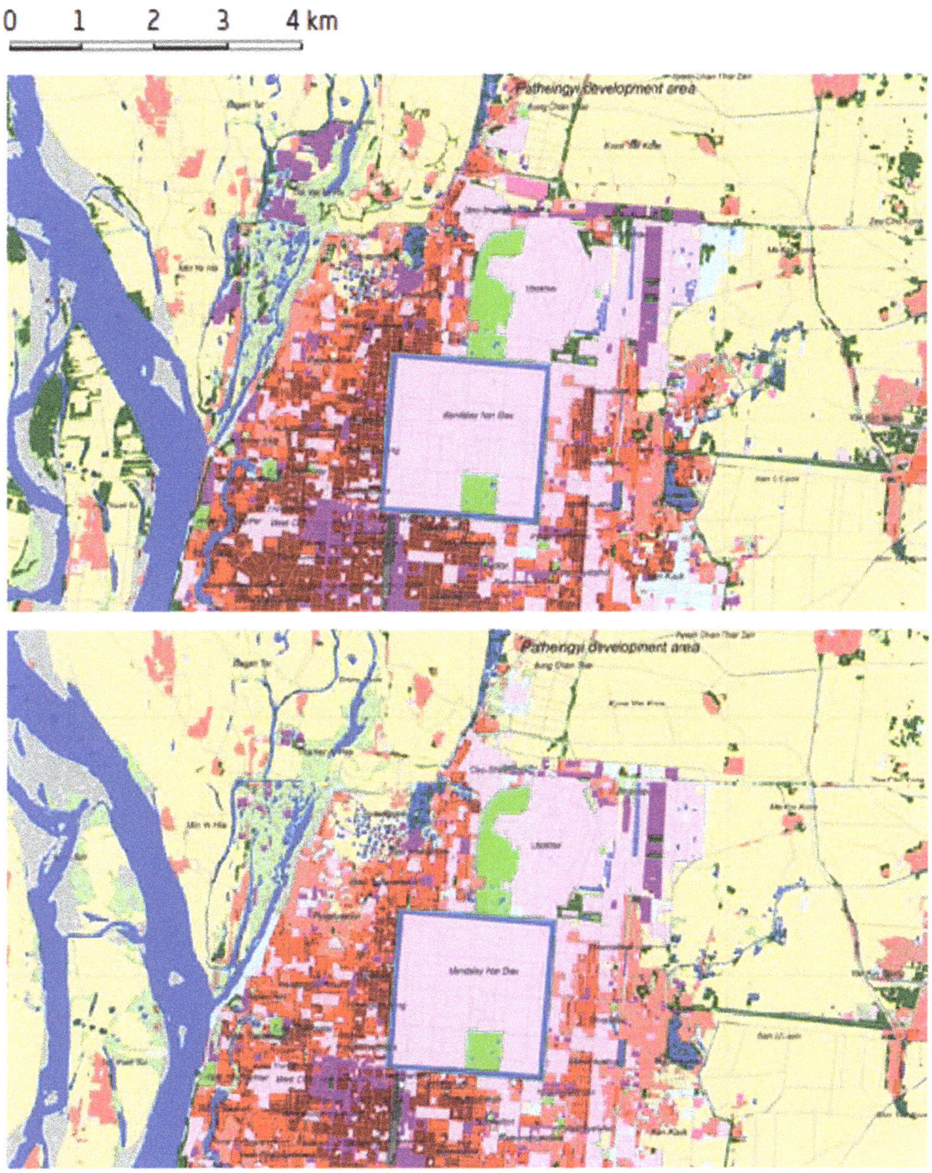

Land use

- Formal very high-density residential (sealing level >80%)
- Formal high-density residential (sealing level 50%–80%)
- Formal low-density residential (sealing level 10%–50%)
- Commercial and industrial
- Nonresidential urban fabric
- Roads, transportation, and associated areas
- Urban greenery
- Construction sites
- Agriculture
- Trees
- Other natural and semi-natural, incl. wetlands
- Vacant land not obviously being prepared for construction
- Bare land
- Water

Note: Top image is for 2014–2015, based on Pléiades 0.5-meter (m) resolution optical imagery; bottom image is the historical 2002 QuickBird, 1.5 m resolution.

Source: GISAT Czech Republic, European Space Agency, reference: ADB. 2017. *Earth Observation for a Transforming Asia and Pacific.* p. 41.

Map 8: Urban Land Use Change Information in Mandalay City, Myanmar

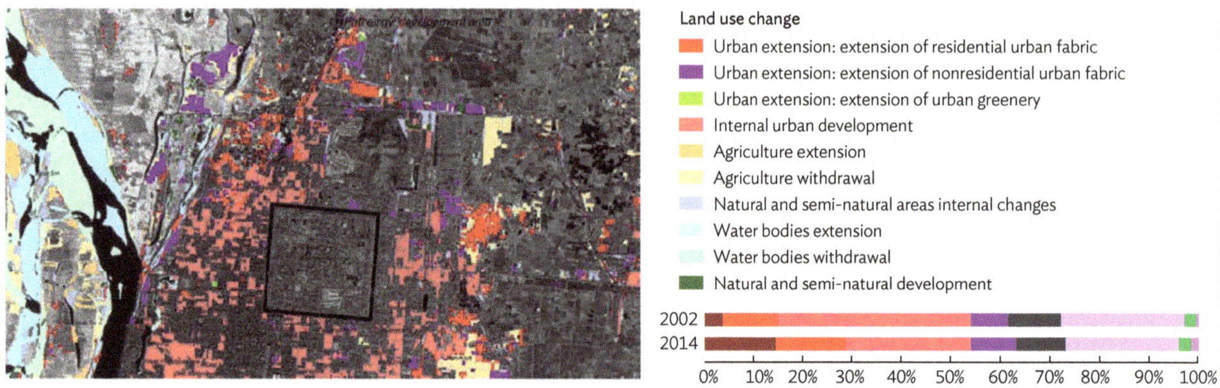

Note: Changes between 2002 and 2014–2015 from the land use information, together with the overall statistics for the land use classes in 2002 and 2014.

Source: GISAT Czech Republic, European Space Agency, reference: ADB. 2017. *Earth Observation for a Transforming Asia and Pacific.* p. 41.

Map 9: Urban Evolution of Yerevan, Armenia, 1995–2020

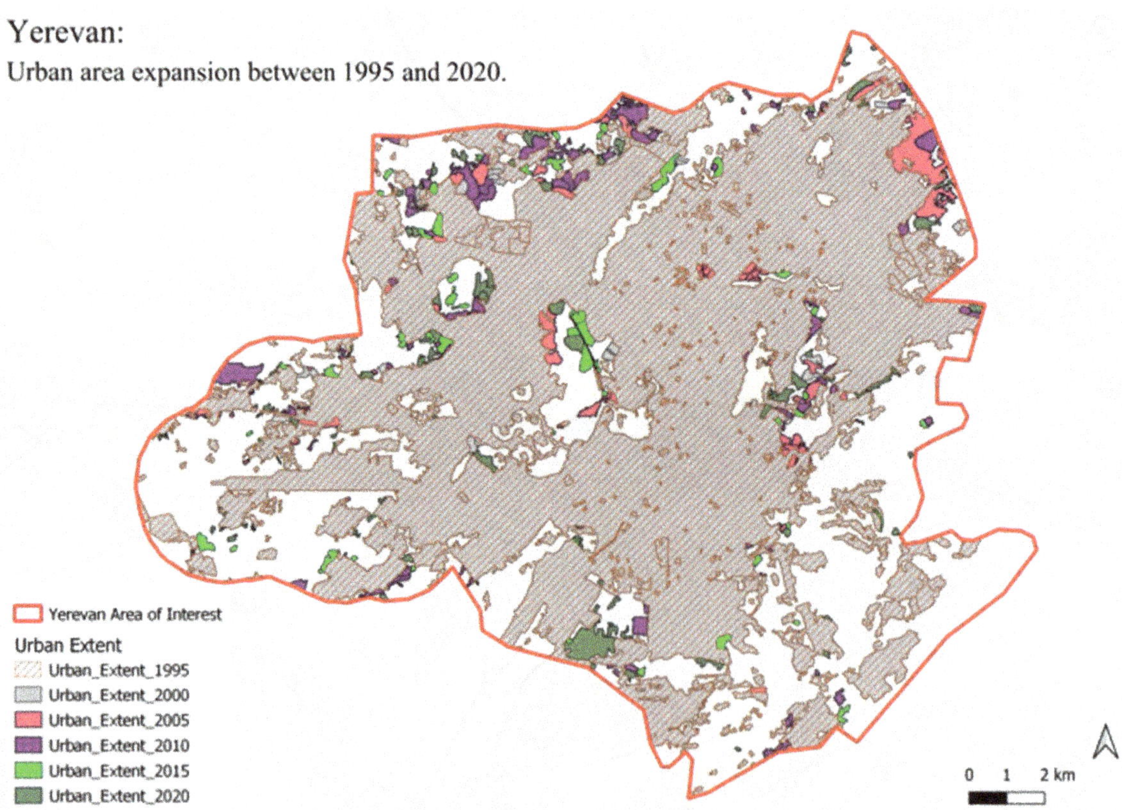

Note: This map was produced from Copernicus Sentinel-2 and LandSat time series optical imagery.

Sources: GeoVille, Austria; CLS, France, ESA, reference: ESA EO Science for Society. EO Clinic: Infrastructure Projects Implementation and Economic Outcomes in Armenia.

Map 10: Zonal Distance Between Building Footprints and Urban Key Facilities in Yerevan, Armenia, 2021

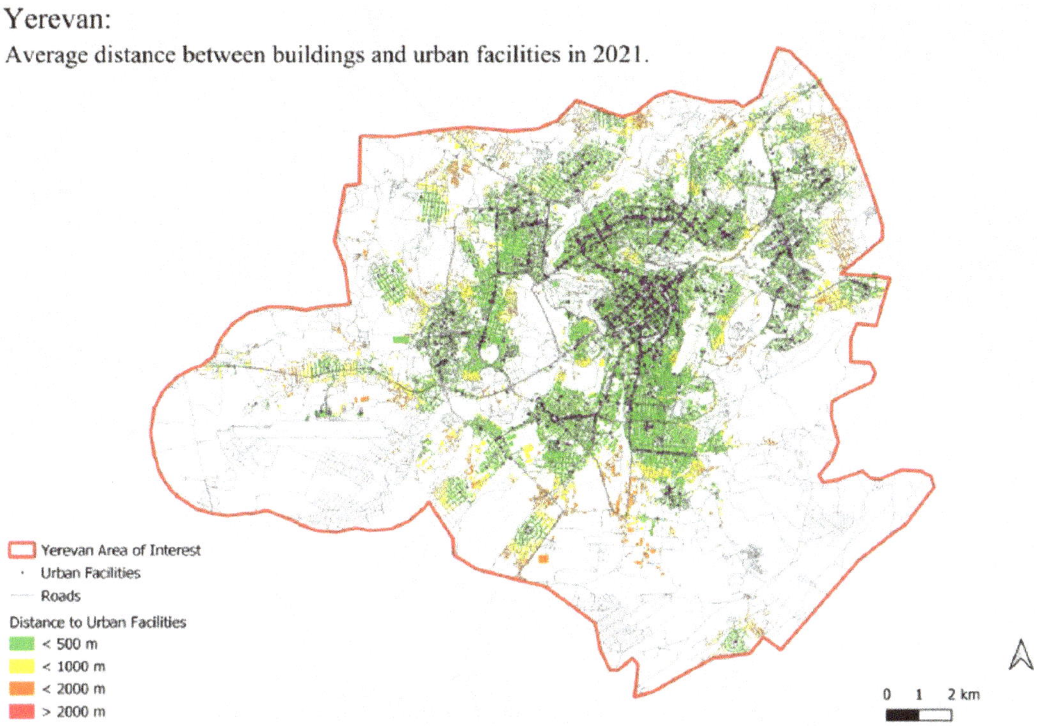

Notes: The average distance metric (color-coded in the map) is a measure of the "reachability" of urban facilities from residential areas. These urban facilities include education centers, health facilities, public transport, urban green areas, and other important facilities. The image shows that most of the buildings in the city are within 500 meters of an economic, administrative, educational, health, or green space facility.

Sources: GeoVille, Austria; CLS, France, ESA, reference: ESA EO Science for Society. EO Clinic: Infrastructure Projects Implementation and Economic Outcomes in Armenia.

Map 11: Expansion of Beijing, People's Republic of China

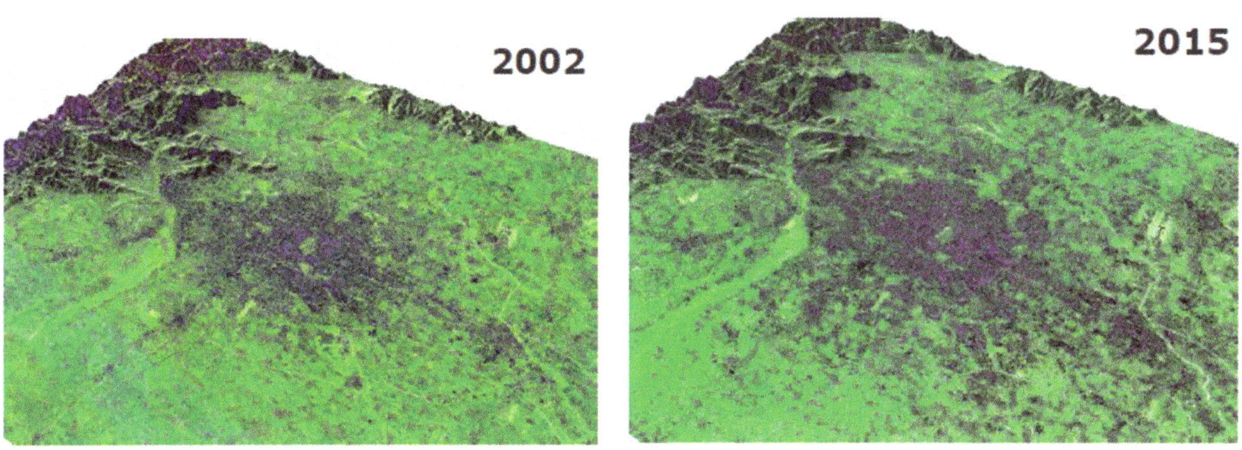

Note: Black areas denote urban extent in 2002 and 2015; derived from Copernicus Sentinel-1A time-series imagery.

Sources: ESA, GAF, reference at European Space Agency. *Earth Observation for Sustainable Development – Product Brochure.* p. 7.

Map 12: Building Height Estimation in Dongying City, People's Republic of China

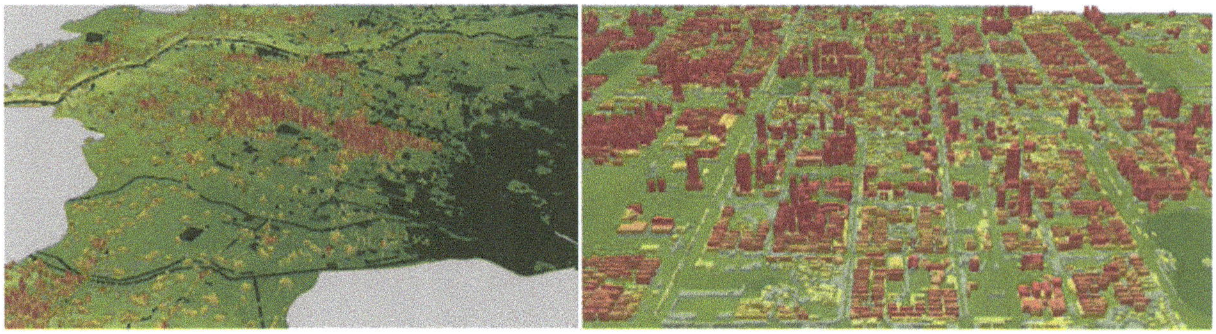

Building Height (m)

- Non Built-up
- 0–3
- 3–6
- 6–9
- 9–12
- 12–15
- 15–18
- >18

Note: The 3-D building model is derived from Pleiades optical imagery at 4-meter resolution.

Source: European Space Agency. *Earth Observation for Sustainable Development – Product Brochure*. p. 27.

Map 13: Informal Settlement Mapping in Jakarta, Indonesia

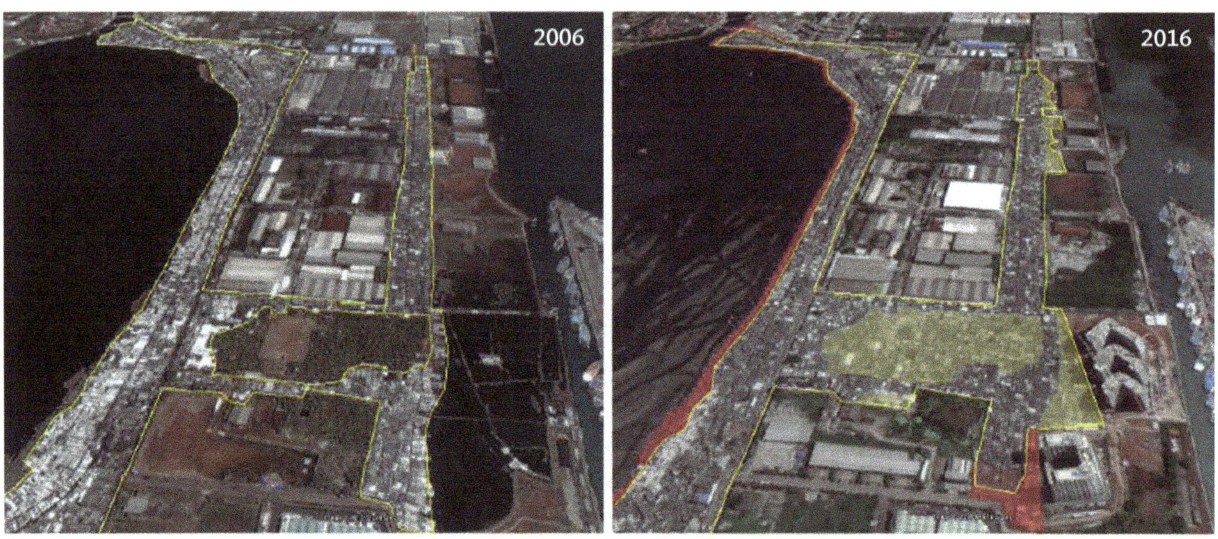

- Slum extent outline
- Slum extension
- Slum reduction

Note: Produced from VHR satellite optical imagery; image on the right shows an analysis of the evolution from 2006 to 2016.

Sources: GISAT Czech Republic, ESA, reference at European Space Agency. *Earth Observation for Sustainable Development – Product Brochure*. p. 10.

- Digital Twins create a digital model of a region or urban area by integrating a range of environmental information layers based on EO together with other data from multiple non-EO sources (e.g., mobile location, census data, aerial surveys, and open geospatial data). The Digital Twin is used for visualization, modeling, and analysis to support regional planning and prioritize development investments. An example is the Geodigital Sumatra with key stakeholders including the ADB Indonesia Resident Mission, BAPPENAS, and ORPA.[125]

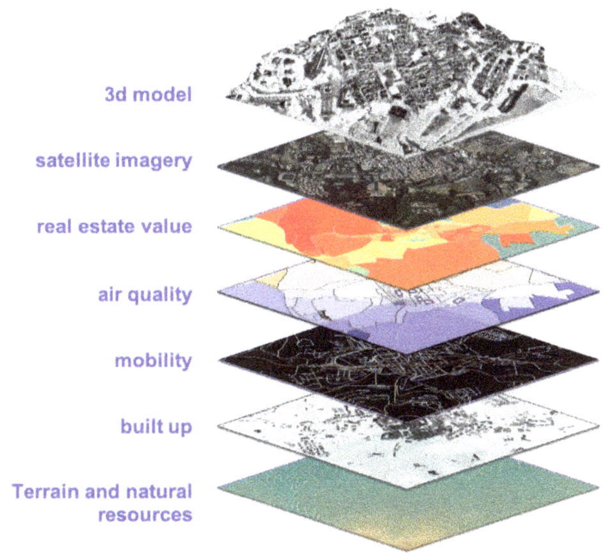

Information layers for Digital Twin City modeling, GeoDigital Sumatra, Indonesia (image by MindEarth, Switzerland).

In summary, how are the land surface type mapping information products that are relevant to the development sector generated from EO satellite data?

Land surface type mapping and monitoring underpins basic environmental and natural resources management, agricultural management, and urban development. As such it is relevant to a wide range of development challenges including:

- raising agricultural productivity at individual farm level, as well as national and regional levels, improving watershed management for rainfed agriculture, and strengthening food security through reducing risk and vulnerability;
- enhancing environmental services and sustainability, identifying the areas prone to land degradation taking into account soil degradation and land use changes, and maintaining biodiversity and ecosystem services; and
- developing sustainable urban growth (energy, transport, fragmentation), including access to clean water, sanitation, electricity; ensuring urban public health (e.g., air pollution); and reducing urban poverty (e.g., informal settlements).

Land surface type mapping and monitoring are directly connected to three of the seven ADB Strategy 2030 operational priorities: *C. Tackling Climate Change, Building Climate and Disaster Resilience, and Enhancing Environmental Sustainability, D. Making Cities More Livable*, and *E. Promoting Rural Development and Food Security.*

[125] Geo Digital Twin. https://geodigitaltwin.org/.

3.2 Earth Observation Use Case 2: Disaster Risk Management and Disaster Risk Reduction

The main EO information content consists of hazard and risk mapping for specific areas within disaster risk management (DRM) (e.g., geohazards, hydrometeorological and climatological hazards). The need for geo-information concerns both *ex ante* activities (before the event) to support understanding of hazards and mitigating risks during preparation phase, and *ex post* activities (after the event), in terms of impact assessment, to support activities associated with recovery, reconstruction, rehabilitation planning, and monitoring and evaluation.

- **Hazard mapping (historical)** involves large-scale spatial and temporal characterization of areas affected by or vulnerable to a hazard (i.e., mainly regional and country-level terrain deformation mapping, flood extent, and frequency mapping). This large-scale hazard mapping is often combined with other bio-geophysical information about the area to characterize and better understand the hazard risk. This additional information can include land surface type (cover) mapping, reference mapping to provide a baseline situation map and asset mapping (see Section 3.1), and Digital Elevation Model for topography and tropical cyclone tracks, etc.

Map 14: Large-Scale Ground Subsidence Mapping, Indonesia

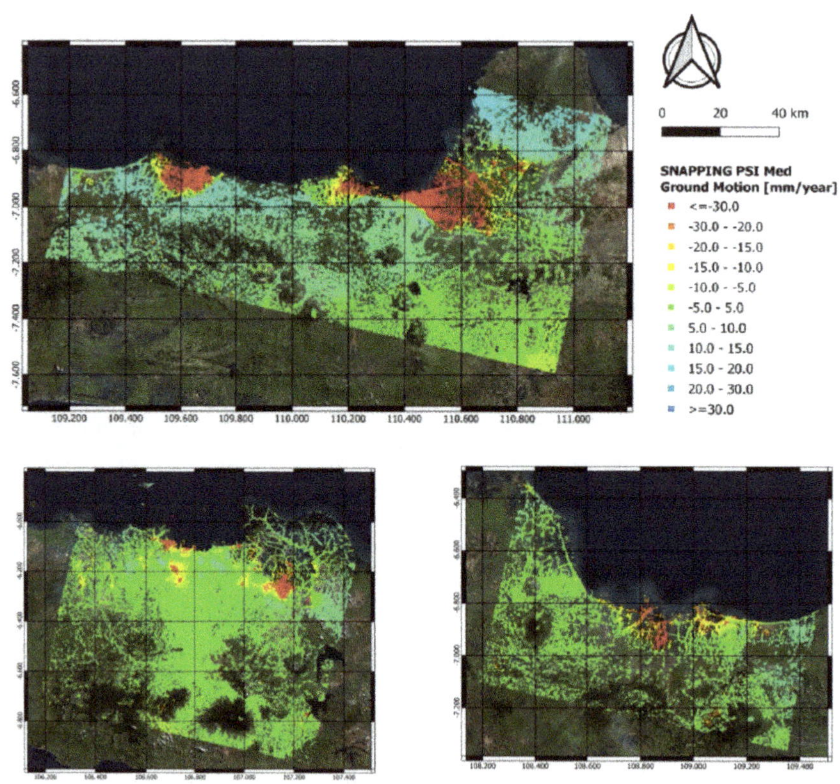

Note: Color scale in millimeters per year using Copernicus Sentinel-1 time series for the regions of (clockwise from top image) Pekalongan and Semarang, Cirebon, and Jakarta.

Source: Terradue, Italy, reference: ADB. 2017. *Earth Observation for a Transforming Asia and Pacific*. p. 2.

Map 15: Large-Scale Flood Extent Mapping (top) and Flood Frequency Mapping (bottom) in the Cirebon Region, Central Java

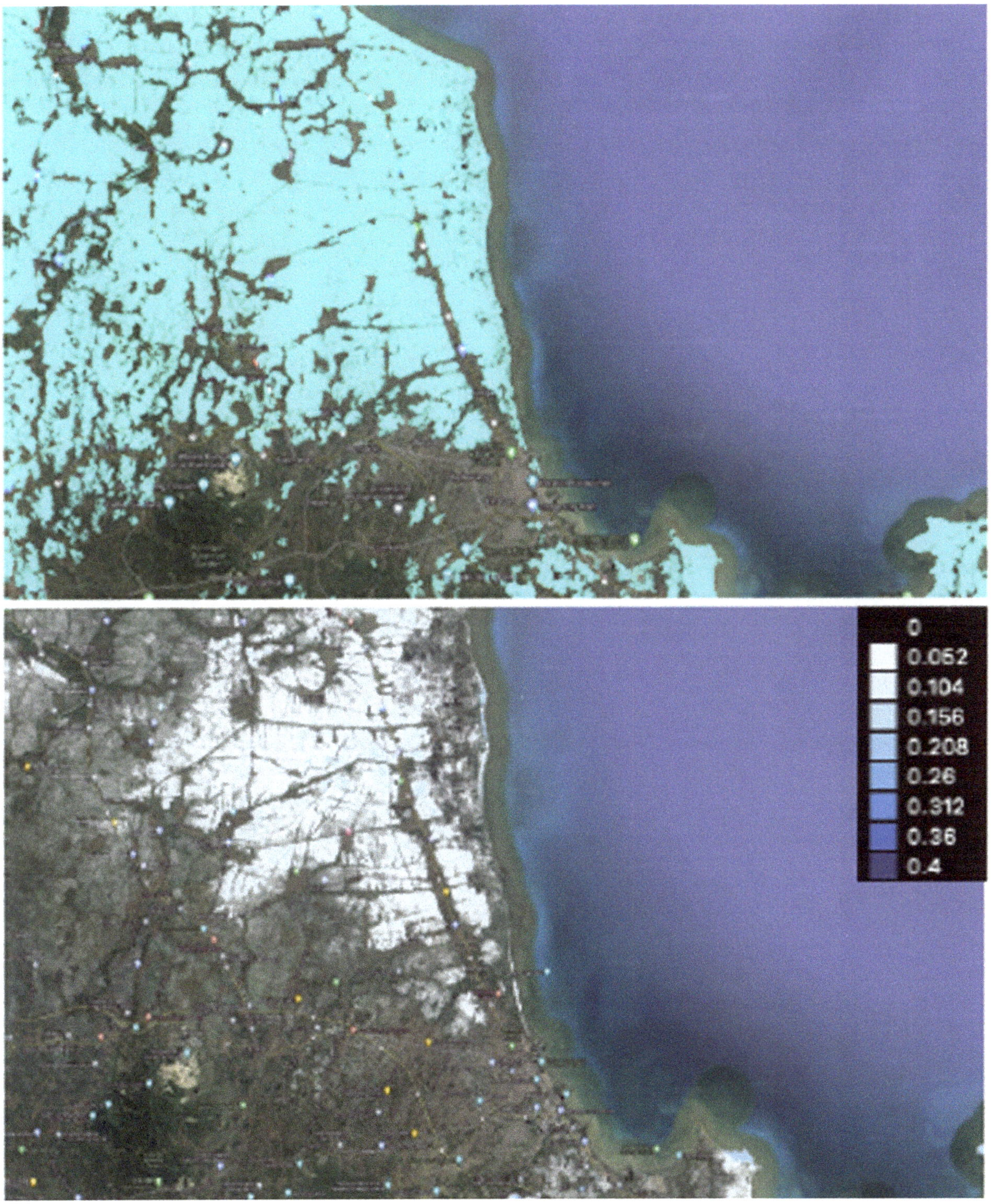

Source: CIMA Research Foundation, Luxembourg Institute of Science and Technology; European Space Agency. Disaster Resilience.

- **Hazard mapping (event-based)** involves detailed and localized measurement and monitoring of areas affected by or vulnerable to a hazard associated with a particular event in time, mainly current terrain motion (e.g., ground subsidence and ground swelling due to earthquakes and seismic events or groundwater extraction); landslide inventories; volcano hazard monitoring; flood event extent; extreme weather events (hurricanes, cyclones, typhoons); fire events; drought events; snow and ice; ocean waves and surges; oil spills; and other events such as technological disasters (e.g., dam failures), diseases, and sandstorms.

  This localized hazard mapping is often combined with other bio-geophysical information about the area to characterize the hazard impact in terms of damage assessment. This additional information can include land surface type (cover) mapping, reference mapping, asset mapping (see Section 3.1), and Digital Elevation Model for topography.

Map 16: Evolution of an Extensive Flood Event in Larkana, Pakistan Due to Extreme Monsoon Rains, June–August 2022

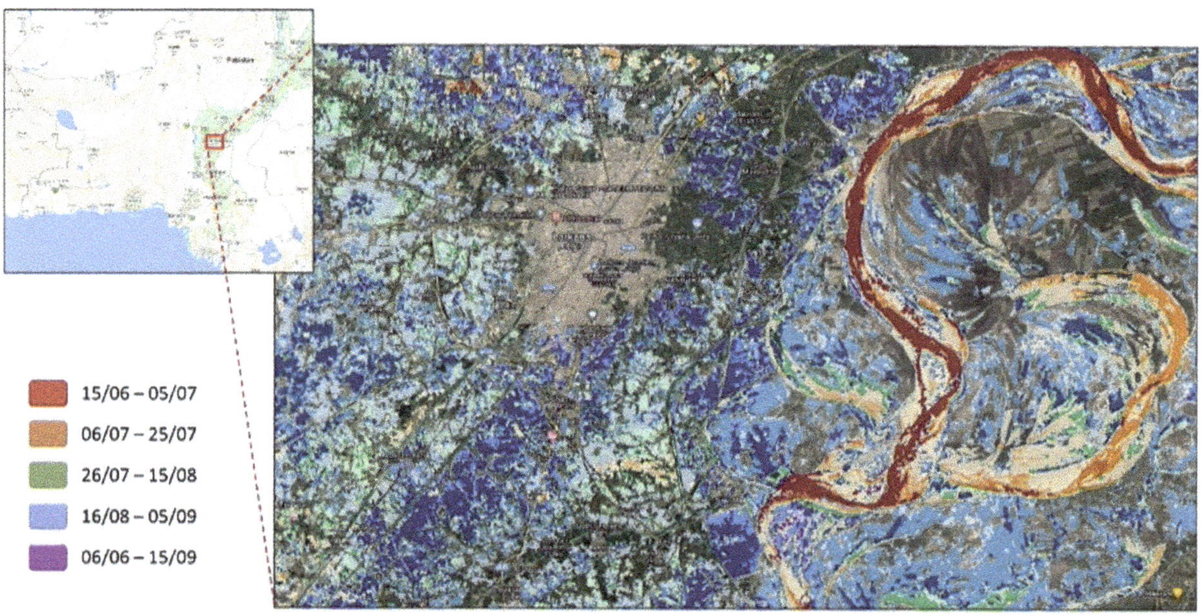

Note: The flood evolution map was produced from Copernicus Sentinel-1 SAR imagery.

Sources: CIMA Research Foundation, Luxembourg Institute of Science and Technology; CIMA Foundation. EO for Pakistan Floods; and M. Belenguer-Plomer, C. Fafet, and R. Rudari. 2022. European Earth Observation providers support ADB in Pakistan floods recovery efforts. 22 November.

Map 17: Maximum Estimate Water Depth in Nasirabad, Larkana, and Kambar, Among the Sindh and Balochistan Regions of Pakistan, 15 June–15 September 2022

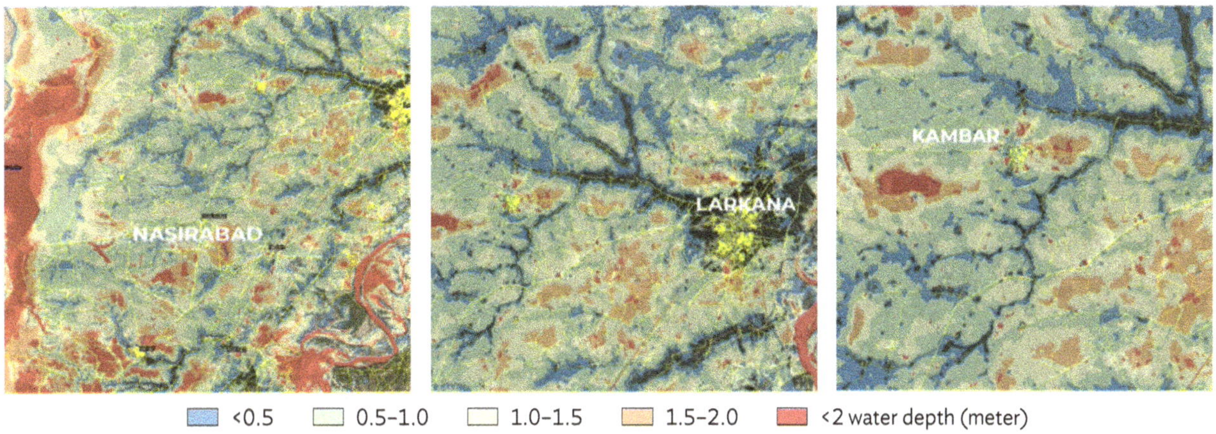

Note: The estimate is performed by combining a flood delineation map captured by the Copernicus Sentinel-1 synthetic aperture radar (SAR) satellite with the FABDEM, a digital elevation model post processed for hydraulic application by the University of Bristol.

Source: CIMA Foundation. EO for Pakistan Floods.

Figure 22: Damage-and-Loss Assessment for the Flood Event in Pakistan, Summer 2022

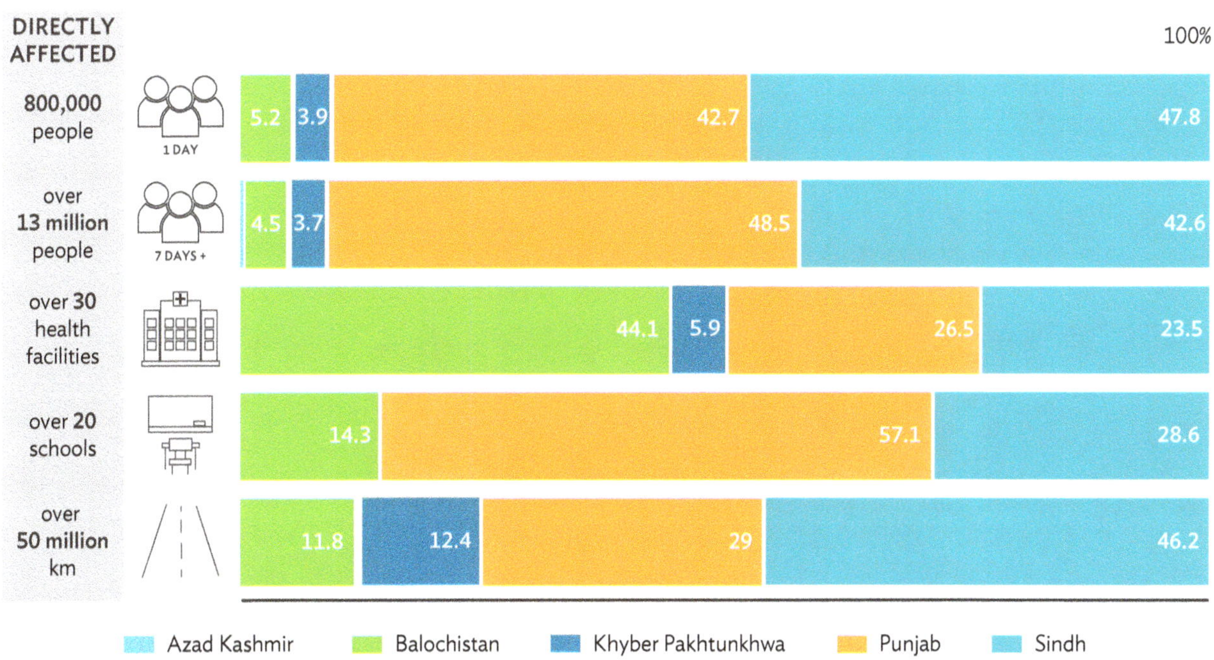

Notes: The most affected populations were those in the Sindh and Punjab regions. In Balochistan, although the population appears to be less affected, the infrastructures, especially the health facilities, were significantly affected.

Source: CIMA Foundation. EO for Pakistan Floods.

Map 18: Example of Land Use Classification Maps for Artificial Surfaces (Transport-Related) and Waterways, near the City of Mendi, Central Papua New Guinea

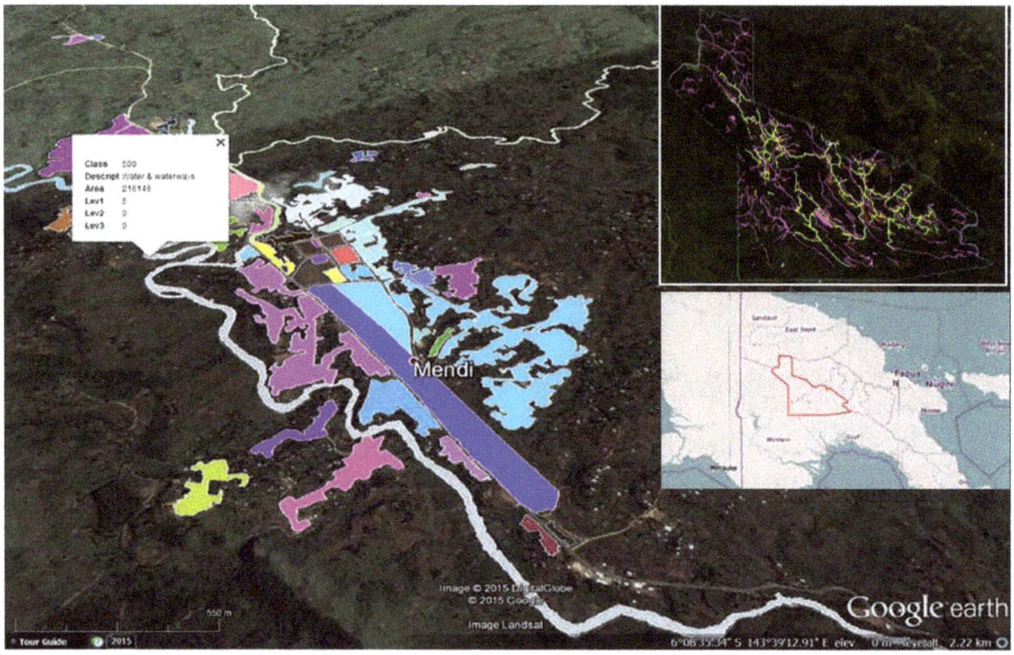

Notes: The insert in top right-hand corner is the extracted road network (yellow) and surface water bodies (pink) for the two provinces of Southern Highlands and Hela located in central Papua New Guinea (provincial boundaries in red in bottom insert). The processing is based mainly on optical RapidEye imagery of 5-meter (m) resolution. Where the optical data was affected by cloud cover, the processing was complemented by RADARSAT-2 (radar, 3 m resolution).

Source: Image from GAP, Planetek, Italy, for European Space Agency and Asian Development Bank, reference: ADB. 2017. *Earth Observation for a Transforming Asia and Pacific*. p. 21.

Map 19: Papua New Guinea: (Right) Large Landslide Observed on RapidEye Image of 5-Meter Resolution (30 September 2012) and (Left) Same Location on a Pan-sharpened Pléiades Image of 0.5-Meter Resolution (4 February 2014)

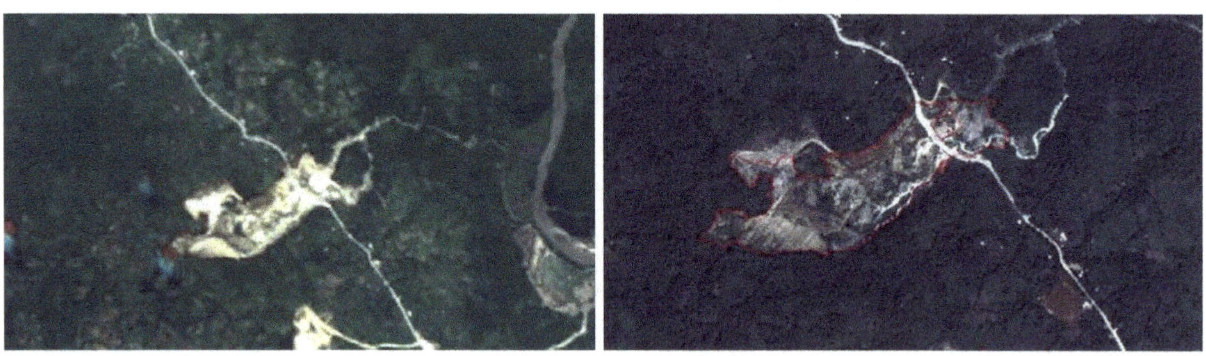

Notes: More detailed geo-morphic features (e.g., head scarp, drainage) are clearly visible on the higher-resolution image (right). The landslide is located along the Spine-road area in the area of interest, where construction of road on a complex topography, with loose materials often piled up along the road, has led to many debris slides and flows.

Source: Images from GAP, Planetek, Italy, for European Space Agency and Asian Development Bank, reference: ADB. 2017. *Earth Observation for a Transforming Asia and Pacific*. p. 23.

Map 20: Example of a Land Surface Displacement Map over the City of Mendi, Papua New Guinea Displayed in Google Earth

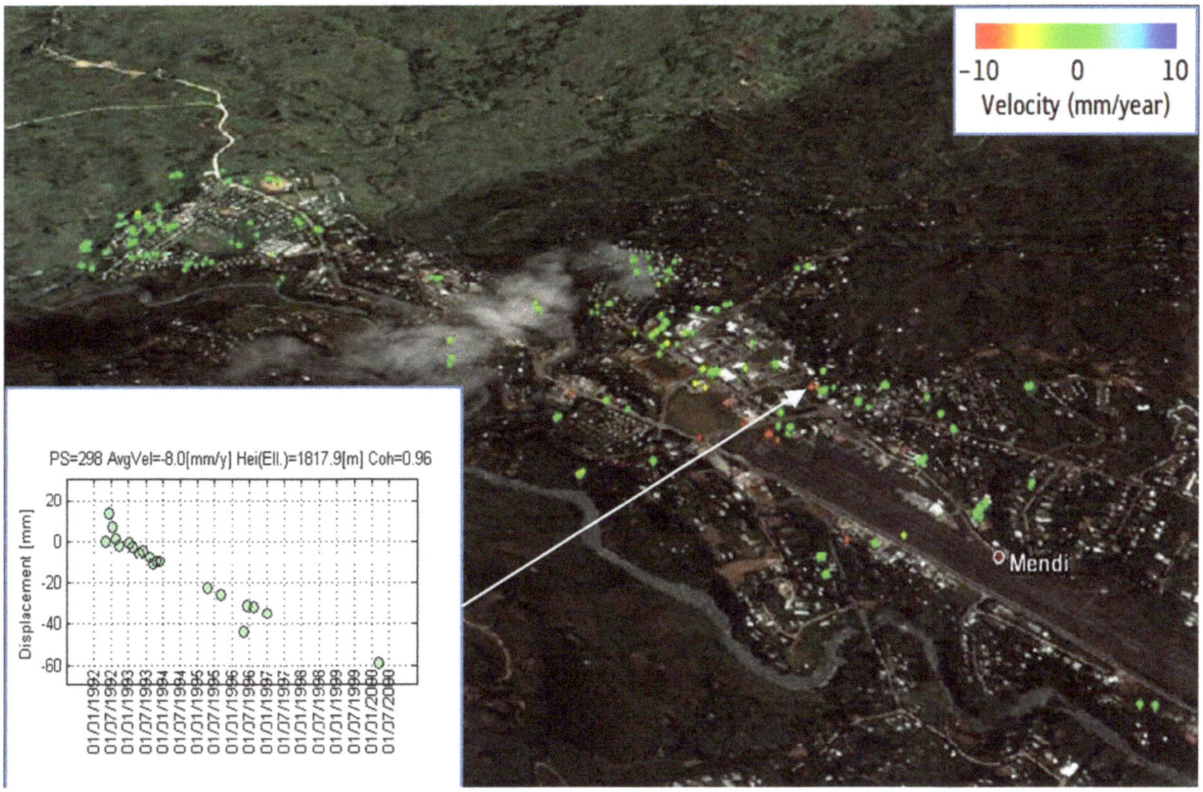

Source: Image from GAP, Planetek, Italy, for European Space Agency and Asian Development Bank, reference: ESA/ADB publication *Earth Observation for a Transforming Asia and Pacific* (March 2017) at https://www.adb.org/publications/earth-observation-transforming-asia-pacific/, p. 23.

The image in Map 20 was produced from ESA ERS-1 radar image data for the period 1992–2000.
For each point (called Persistent Scatterers), the color represents the estimated average land displacement velocity, as indicated in the color legend, top right.

However, this average velocity is only a fraction of the information that can be retrieved. For each Persistent Scatterers point, the full temporal evolution of the displacement is measured, as indicated by the insert figure on the bottom left.

For that particular point, a mostly linear trend of –8 millimeters/year could be observed. Such land surface displacement maps help assess the risk and vulnerability of the area to potential landslide events.

The purple points in Map 21 show significant motion of a potentially ice-cored scree slope in the upper central part of the watershed. The map was produced from Copernicus Sentinel 1-A data during the time period 2015–2022.

This terrain motion information was used to guide the design of an in-situ early warning system to capture potential natural hazard processes that occur in the area, such as landslides, floods, and glacial lake outburst floods.

Map 21: Terrain Motion in Several Areas Across the Melamchi Watershed in Central Nepal, North of Kathmandu, 2015–2022

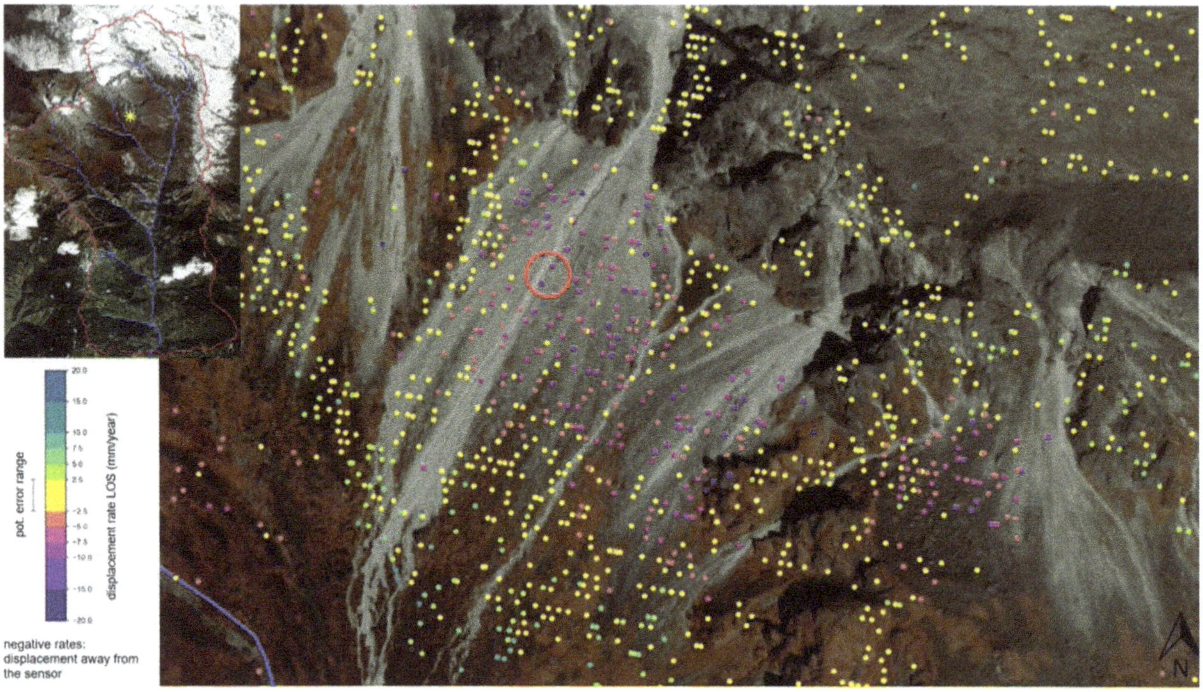

Source: Image from Gisat, Czech Republic, for European Space Agency and Asian Development Bank, reference ADB Report Nr. 1422063.4a, Conceptual Design of an Early Warning System: Melamchi Water Supply Project, Nepal, GEOTEST AG, July 2023.

- **Risk mapping** involves combining hazard exposure and vulnerability via mapping and modeling to support risk mitigation and prevention as well as preparedness. This can include geo-hazards, hydrometeorological hazards, and climatological hazards.

Map 22: Example of Infrastructure Risk Map in Jakarta, Indonesia

Notes: (Left) Ground motion histories combined with building footprints from Open Street Map to produce a detailed Buildings Stability Indicator (red, yellow, green, not applicable). The processing techniques take into account motion histories of individual buildings as well as nearby areas.

Source: Image from Planetek Italia, for European Space Agency and Asian Development Bank, reference https://gda.esa.int/thematic-area/disaster-resilience/#supported.

Note that this Flood Risk Index is processed as a function of maximum historical flood water depth, population density, and building typology.

Map 23: Detailed Map of Buildings Affected in the Flood Risk Zone, Jakarta, Indonesia

Source: Image from Planetek Italia, for European Space Agency and Asian Development Bank. https://gda.esa.int/thematic-area/disaster-resilience/#supported.

The graphic below is a statistical analysis of the buildings affected in the flood risk zone.

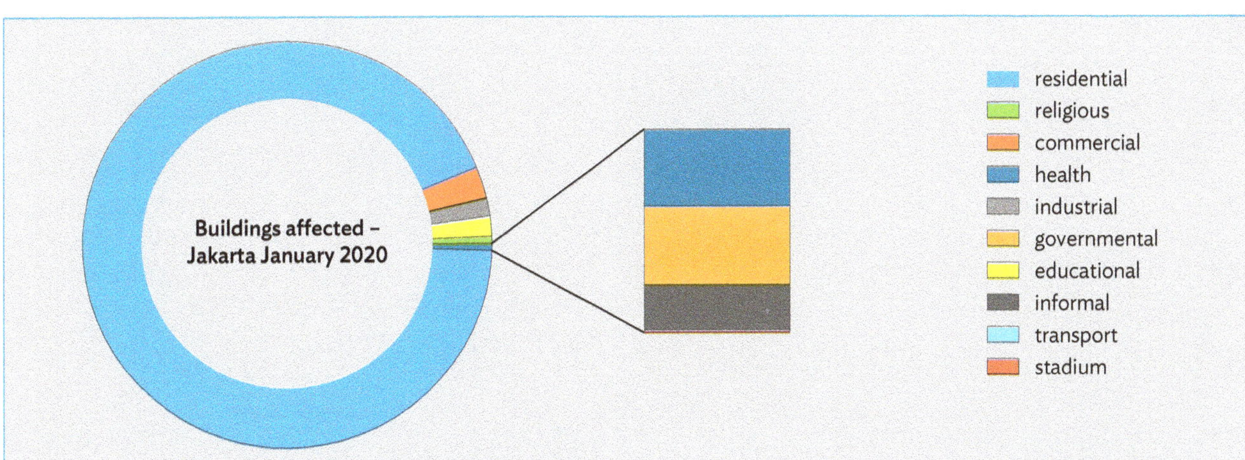

Source: Image from Planetek Italia, for European Space Agency and Asian Development Bank. https://gda.esa.int/thematic-area/disaster-resilience/#supported.

Note this flood frequency history is generated by integrating radar data (ALOS PALSAR) and optical imagery (SPOT 4/5) at 10–20 m resolution, measuring the flood situation during the peak monsoon flood inundation periods for 2003–2009.

Map 24: Example of Flood Frequency Map for Mandalay City Area, Myanmar

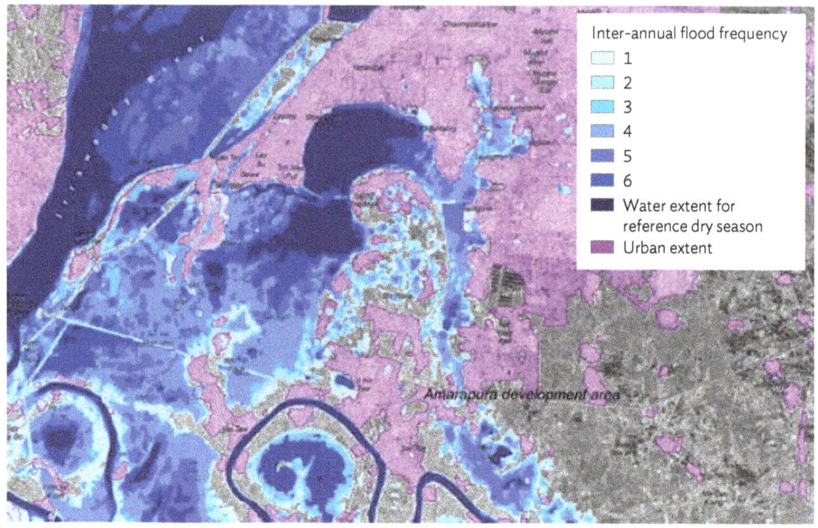

Source: Image from GISAT Czech Republic, for European Space Agency and Asian Development Bank, reference at ESA. *EO4SD Product Brochure*. p. 29.

Note, this Flood Risk Index map was generated for potentially affected built-up areas based on flood inundation extent and depth, and detailed digital terrain model from triple stereo processing of SPOT 7 imagery which served as input to the modeling.[126]

Map 25: Flood Risk Index Map for Mandalay City Area, Myanmar

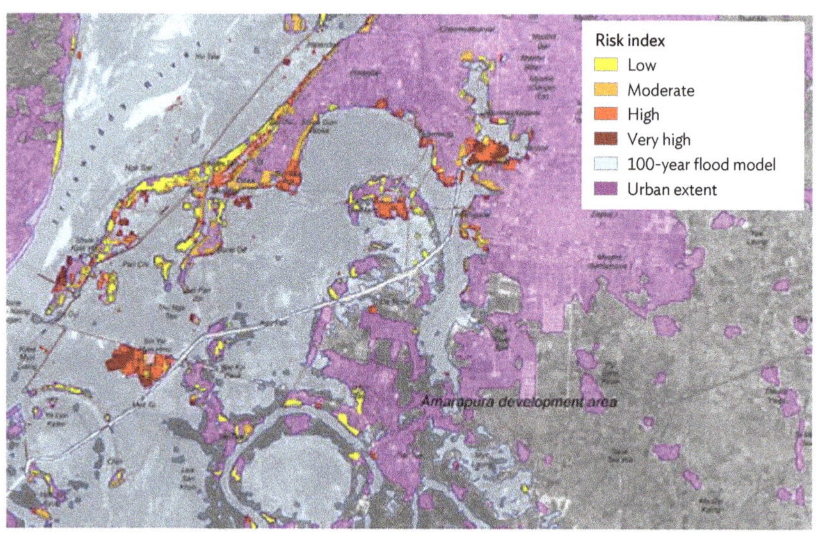

Source: Image from GISAT Czech Republic, for European Space Agency and Asian Development Bank, reference at ESA. *EO4SD Product Brochure*. p. 29.

[126] p. 40; Urban Services Improvement in Myanmar.

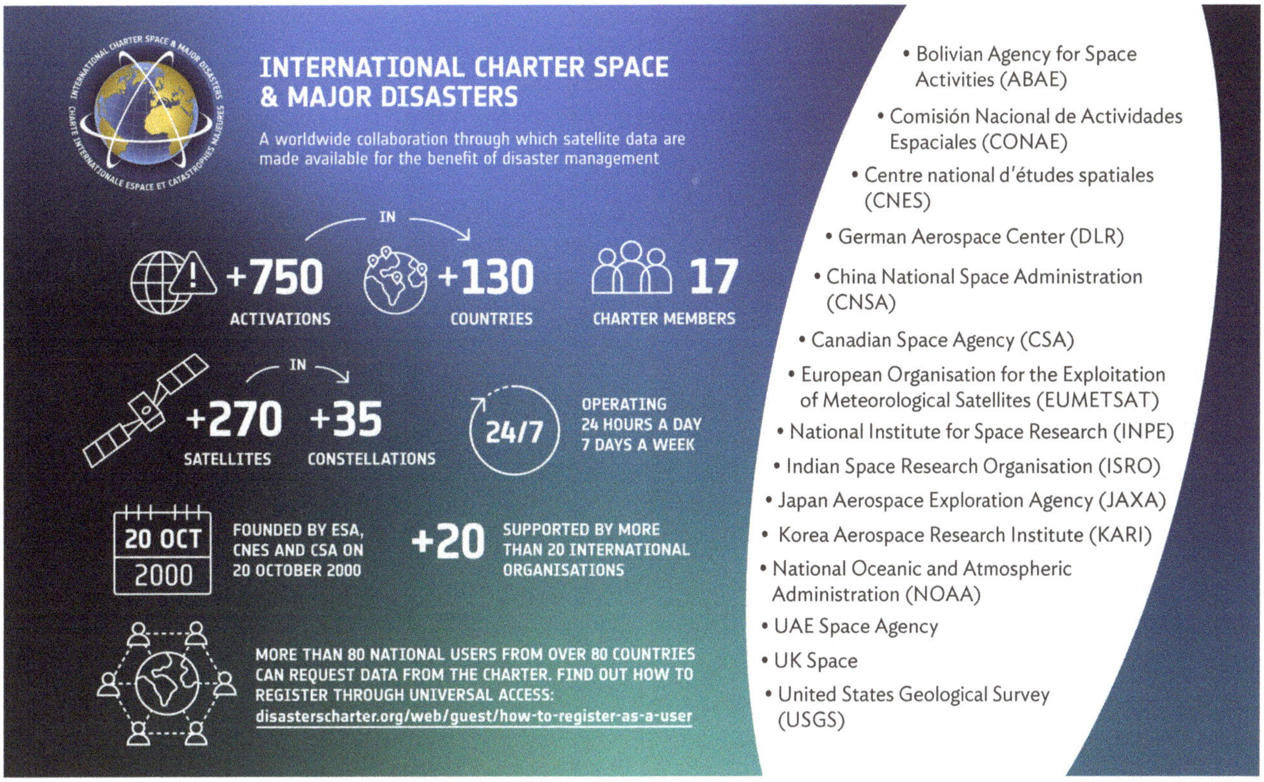

In addition to the specific ADB use cases given above, many more working examples of satellite EO information being used in the context of disaster risk management and monitoring can be found at the International Space and Major Disasters web portal.[127]

The International Charter (or the Charter) is a worldwide collaboration (133 countries, 17 members, 270 contributing satellites) through which satellite EO data are made available for the benefit of natural hazard events management.

By combining EO assets from different space agencies, the Charter allows resources and expertise to be coordinated for rapid response actions by civil protection authorities and the international humanitarian community to major disaster situations.

To date, the Charter has been activated on 852 occasions since it was founded on 20 October 2000 by the ESA, the French national space agency (Centre National d'Etudes Spatiales), and the Canadian Space Agency. The activation frequency is, on average, roughly once per 10 days with 50% of activations being flood events.

This unique initiative can activate agencies around the world and benefit from their know-how and their satellites through a single access point that operates 24-7 and at no cost to the user.

A few examples of the use of EO for additional types of disaster events in the 'Risk Mapping' category from recent Charter activations are given in the following maps.

[127] The International Charter: Space and Major Disasters.

Map 26: Example of Luzon Mw 7.0 Earthquake, 27 July 2022, Philippines

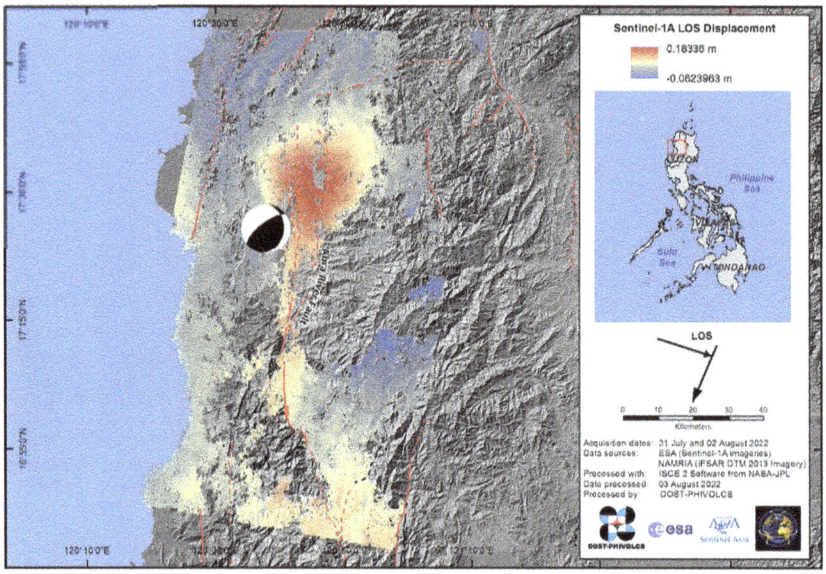

Line-of-sight surface displacement map produced from Copernicus Sentinel-1 SAR imagery showing a north–south deformation pattern along the Abra river fault with up to approximately 20 centimeters measured displacement.

Source: Image from Copernicus data, processed by Department of Science and Technology, Philippine Institute of Volcanology and Seismology (DOST-PHIVOLCS), Charter activation 765, 27 July 2022 at https://disasterscharter.org/activations/earthquake-in-philippines-activation-765-.

Map 27: Example of a Potential Oil Spill Event Off the Coast of Naujan, Oriental Mindoro, Philippines, March 2023

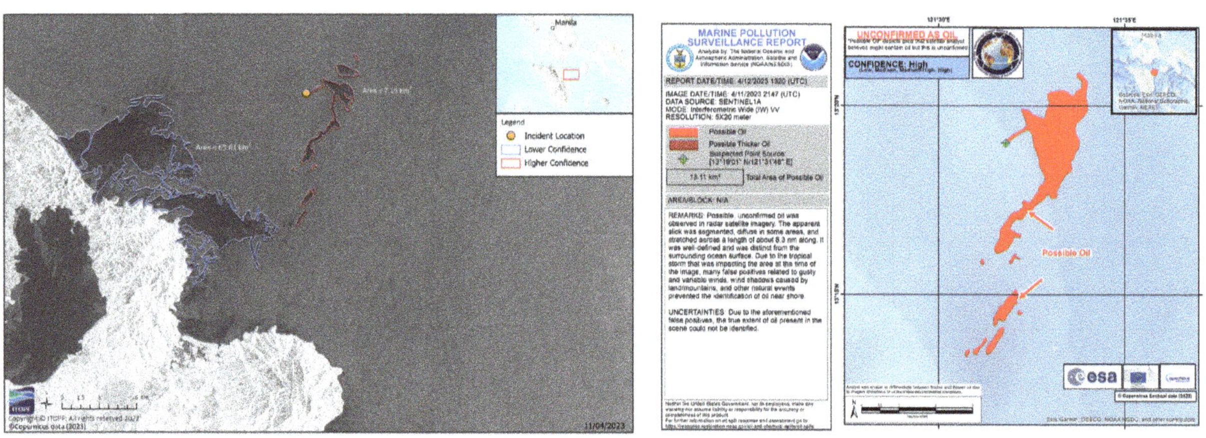

Sources: Left image from the US National Oceanic and Atmospheric Administration, Satellite and Information Service; right image from analysis by International Tanker Owners Pollution Federation, European Space Agency, Philippine Space Agency (PhilSA), Charter activation 803 of 2 March 2023 at https://disasterscharter.org/activations/oil-spill-in-philippines-activation-807-.

On 28 February 2023, the Philippine Coast Guard reported a fuel leak from the oil tanker MT Princess Empress that capsized earlier in the day (image produced from Copernicus Sentinel-1A SAR image data acquired on 11 April 2023).

Map 28: Example of Coastal Damage Mapping Following a Tsunami That Hit Coastal Towns in Indonesia's Sunda Strait on 23 December 2018

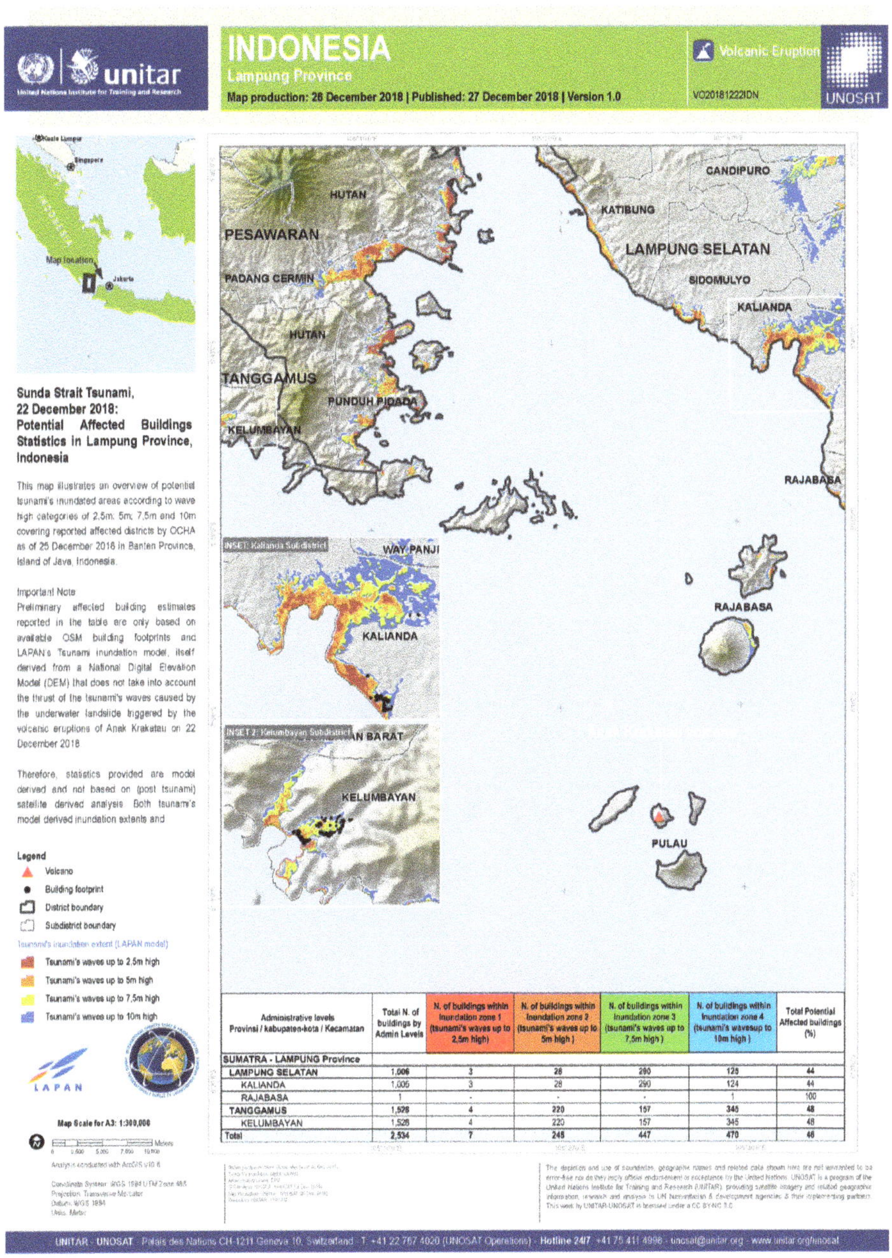

Source: Map produced by UNITAR and UNOSAT on behalf of Indonesian National Institute of Aeronautics and Space (LAPAN), Charter activation 594, 23 December 2018 at https://disasterscharter.org/activations/ocean-wave-in-indonesia-activation-594-.

The tsunami killed at least 228, injured around 1,000, and displaced about 12,000 people. The giant waves struck without warning after an undersea landslide from the Anak Krakatau volcano caused the surges. The color coding is for four separate inundation zones in Lampung province, corresponding to wave heights of 2.5 m (red), 5 m (orange), 7.5 m (green), and 10 m (blue).

Map 29: Example Map of Surface Snow Extent (orange, 11–20 January 2020; red, 18–20 January 2020) in Baluchistan Province, Pakistan

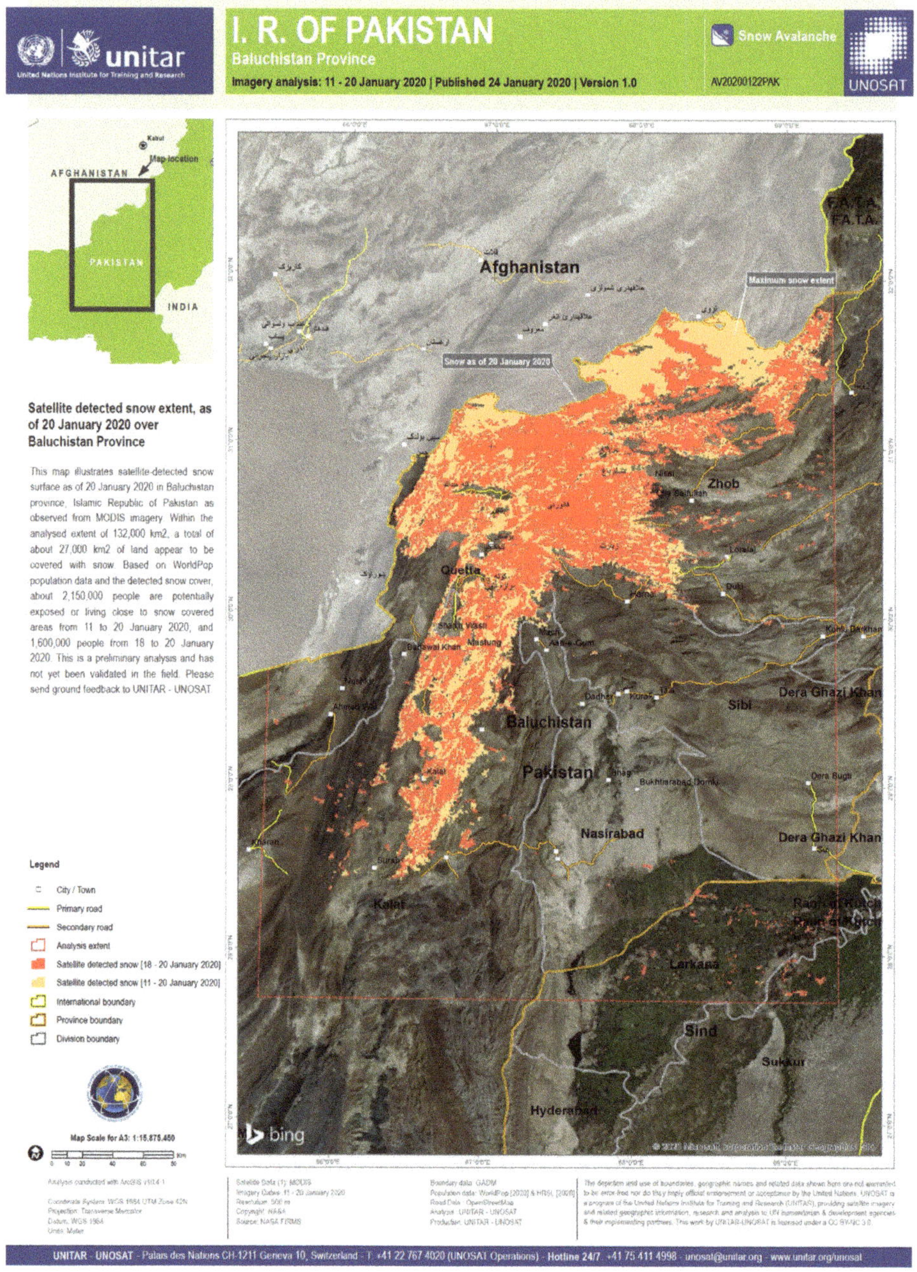

Source: Map produced by UNITAR and UNOSAT, Charter activation 641, 1 Jan 2020 at https://disasterscharter.org/activations/snow-hazard-in-pakistan-activation-641-.

The snow extent map is derived from MODIS satellite imagery. Over 1,000 homes were damaged, and based on WorldPop population data and the detected snow extent, about 2.15 million people were exposed or living close to the snow areas.

Map 30: Example of a Volcanic Ash Plume Map (orange shaded area) from the Eruption of Bulusan Volcano on 12 June 2022 in the Province of Sorsogon, Philippines

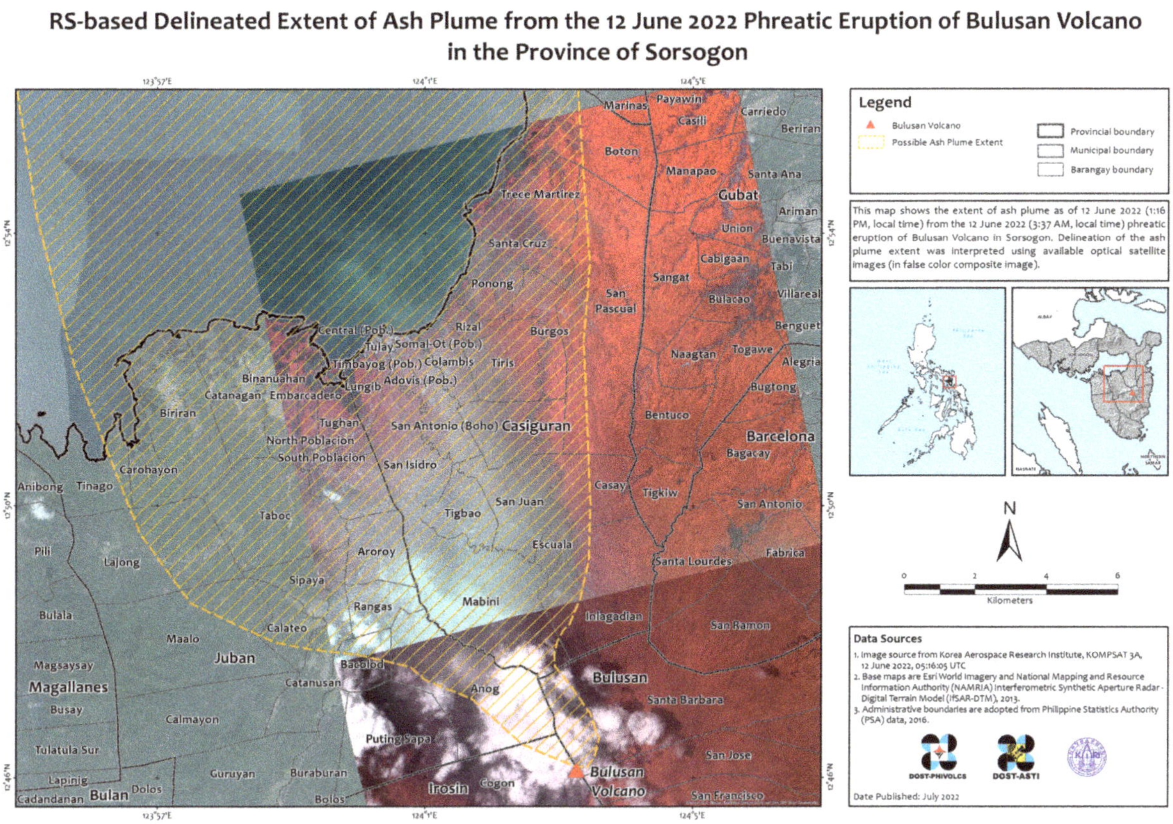

Sources: Map produced by Department of Science and Technology, Philippine Institute of Volcanology and Seismology (DOST-PHIVOLCS), satellite imagery from the Korea Aerospace Research Institute KOMPSAT-3A, Charter activation 761, 13 June 2020 at https://disasterscharter.org/activations/volcano-in-philippines-activation-761-.

An eruption of Mount Bulusan (Sorsogon Province, central Philippines) occurred at 19.37 UTC (3.37 local time) on 11 June 2020.

The ash plume reached up to 500 m high above the crater and the eruption lasted approximately 18 minutes.

The plumes evolved into a long streak of ash extending to the northwest, and the ash fall covered 18 district areas located in the municipalities of Casiguaran, Juban, and Magallanes.

Authorities reported over 400 people were evacuated and more than 15,000 were affected due to the first phreatic eruption that occurred on 5 June.

The image below was captured following the passage of tropical cyclone GITA-16 on 15 February 2018. Analysis identified potentially 637 buildings damaged, which represents 27% of the total buildings analyzed—showing a severe damage overall (satellite imagery by Digital Globe WorldView-3 [pre-event, 12 January 2018] and Airbus Space and Defense/CNES Pleiades [post-event, 15 February 2018]).

Map 31: Example of a Damage Assessment Map of the Eastern Part of Nuku'alofa, Tongatapu Island, Tonga Following Cyclone Passage of February 2023

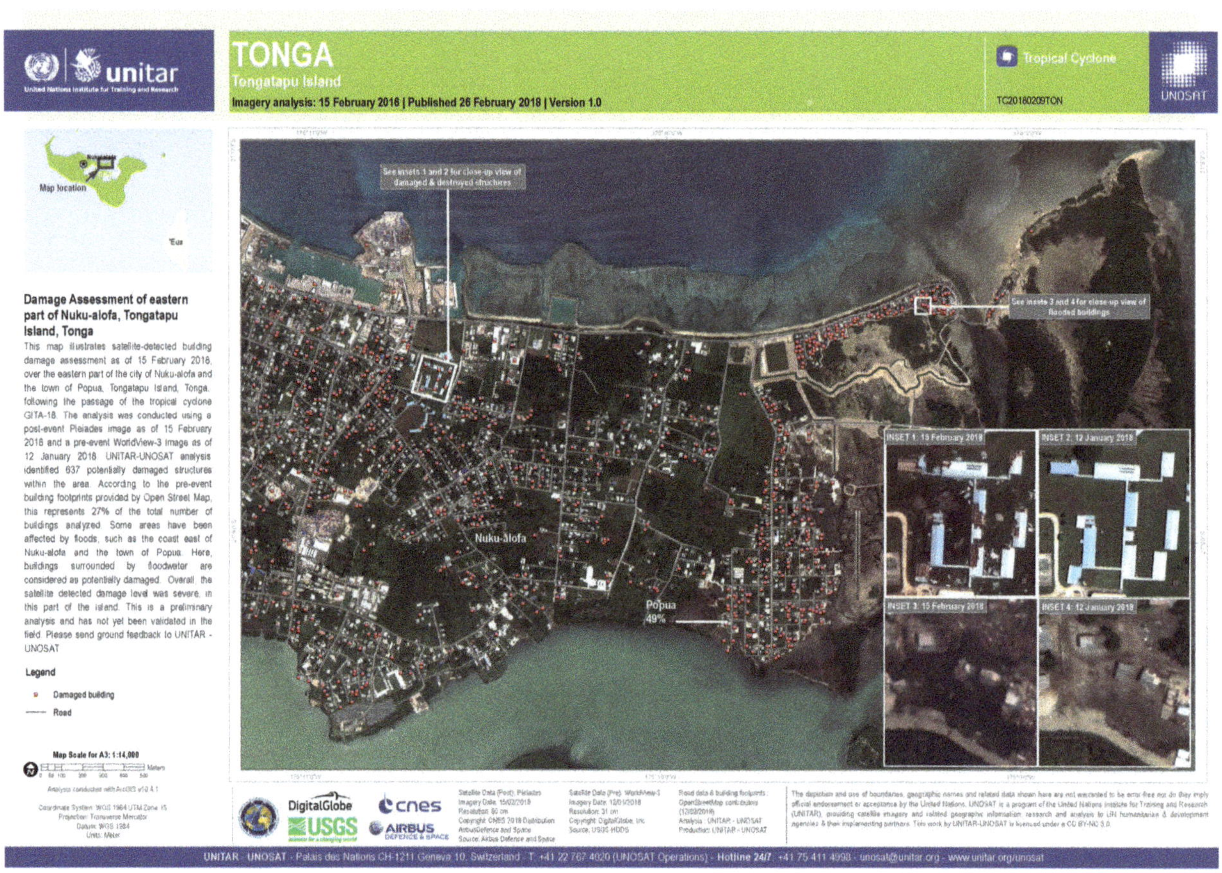

Source: Map produced by UNITAR and UNOSAT. https://reliefweb.int/map/tonga/damage-assessment-eastern-part-nuku-alofa-tongatapu-island-tonga-imagery-analysis-15.

In addition to the International Charter, the Copernicus Emergency Management Service (CEMS) supports the actors involved in the management of natural or human-induced hazard events by providing geospatial data and images for better-informed decision-making.[128] CEMS products are created using EO satellite, in-situ (non-space) and model data.

[128] Copernicus Emergency Management Service (CEMS).

CEMS systematically and continuously monitors areas in Europe and around the globe to identify signs of an impending natural or human-made hazard event, or to see evidence of an occurrence in real time. In these circumstances, CEMS immediately notifies the relevant national authorities about the situation.

CEMS can also be activated through an on-demand mechanism, and can provide maps, time sequences, and other relevant information to better manage disaster risks.

In summary, how are these types of disaster-related environmental information products generated from EO satellite data relevant to the development sector?

Natural hazard events cause human harm, environmental damage, and economic losses.

Understanding disaster risk in all its dimensions of vulnerability, exposure of persons and assets, hazard characteristics, and impact on the environment is a priority of the Sendai Framework for Disaster Risk Reduction (2015–2030).[129] The Sendai framework was adopted at the Third UN World Conference on Disaster Risk Reduction in Sendai, Japan on 18 March 2015. It aims to achieve substantial reduction of disaster risk and the consequent losses in lives, livelihoods and health of people and communities, and in the economic, physical, social, cultural, and environmental assets of businesses and countries.

Asia and the Pacific is prone to many types of natural hazard events including geohazards like earthquakes and volcanoes; hydrometeorological hazards like floods, hurricanes, tropical storms, and storm surges; and climatological events like droughts, heat waves, and wildfires. The impact of these events should be addressed not only by responding after episodes but also by enhancing prevention and preparedness.

EO can contribute to tackling most of these natural hazard types efficiently by providing hazard mapping; supporting services for the assessment of exposure, vulnerability, and risk; and monitoring the reconstruction.

The impact of such disasters on lives and economy is of prime importance to society, especially for developing countries, where mortality and economic losses are disproportionately high and where development achievements can be threatened. ADB plays a significant role in the developing member countries through direct cooperation with national disaster authorities in preventing and mitigating the adverse effects of disasters, as well as in fostering sustainable development.

Disaster risk management and disaster risk reduction are directly connected to three of the seven ADB Strategy 2030 operational priorities: C. *Tackling Climate Change, Building Climate and Disaster Resilience, and Enhancing, Environmental Sustainability*, D. *Making Cities More Livable*, and E. *Promoting Rural Development and Food Security*.

[129] United Nations Office for Disaster Risk Reduction (UNDRR). 2015. *Sendai Framework for Disaster Risk Reduction 2015–2030*.

3.3 Earth Observation Use Case 3: Climate Resilience and Proofing

To support the work of the United Nations Framework Convention on Climate Change (UNFCCC) and the Intergovernmental Panel on Climate Change (IPCC), the Global Climate Observing System (GCOS), sponsored by the World Meteorological Organization, created a list of key parameters of the Earth system, known as the Essential Climate Variables (ECVs).[130] This information is critical for informed decision-making in climate variability mitigation and adaptation.

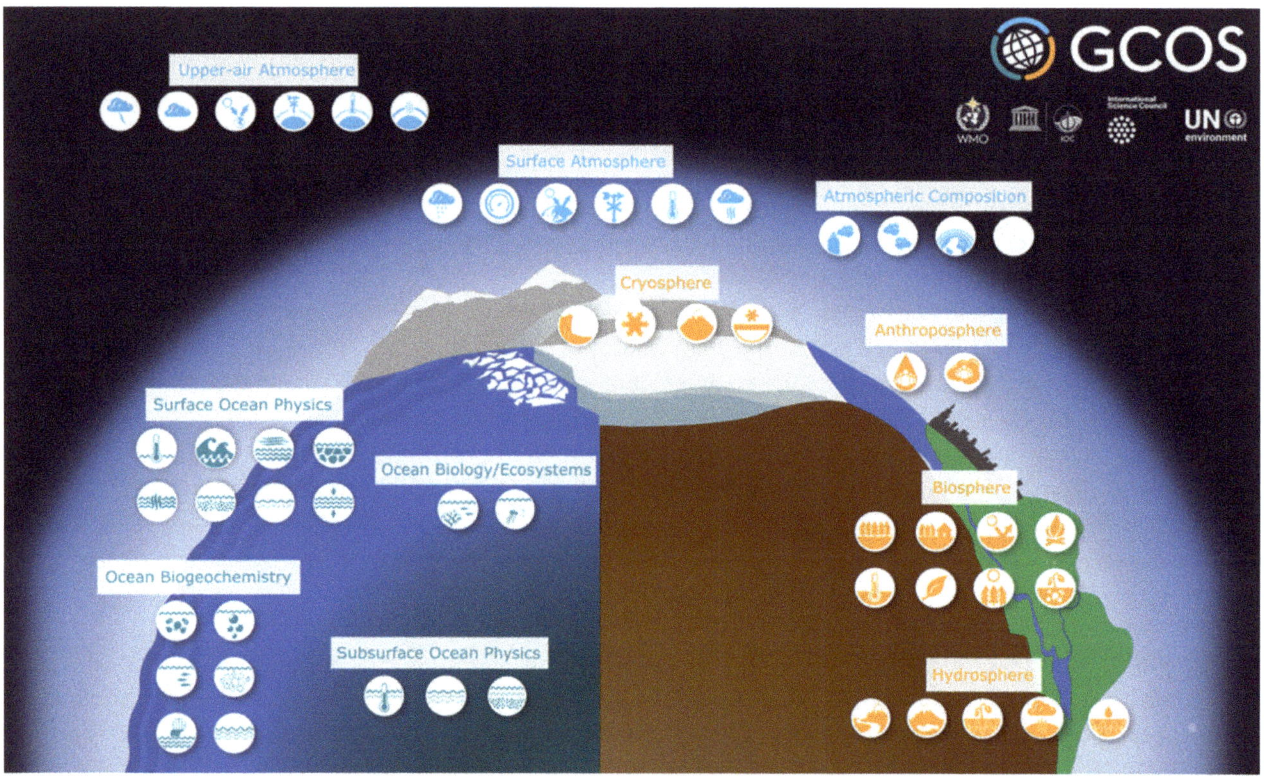

▲ Global Climate Observing System Essential Climate Variables (image from GCOS).

GCOS have specified 55 ECVs which are physical, chemical, or biological variables to meet the climate information needs of the scientific community and provide the necessary factual evidence to understand how our climate is evolving.

EO satellites can significantly contribute to more than half of the ECVs, making EO a key component of climate monitoring. To produce datasets that cover a sufficient timespan and that allow scientists to identify changes from natural climate system variability (i.e., normally 30 years or more), satellite data can be merged with both historical archive and ongoing satellite missions. There are a number of key international initiatives that are designed to deliver EO-based climate information to decision-makers.

130 Global Climate Observing System (GCOS). Essential Climate Variables.

Map 32: Global Surface Air Temperature Anomaly for 2023 Relative to the Average for the 1991–2020 Reference Period

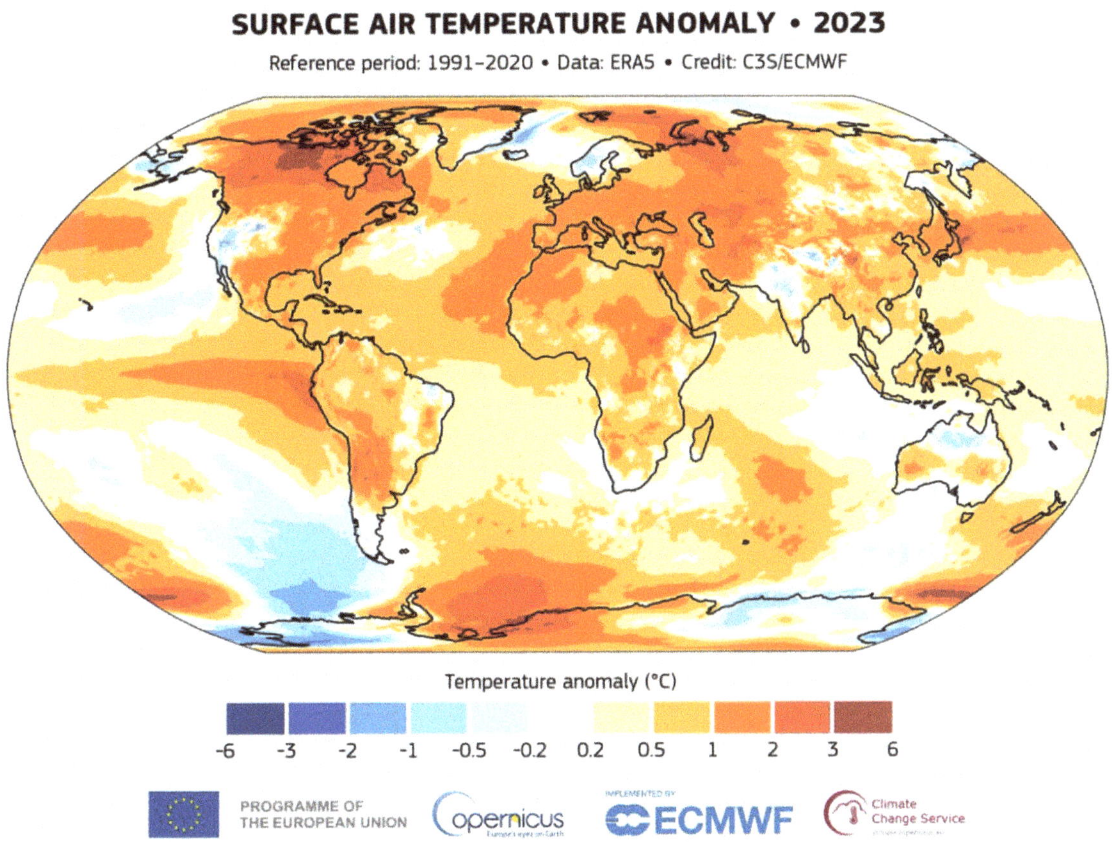

Sources: Data from ERA5; image from the Copernicus Climate Change Service (C3S) and European Centre for Medium range Weather Forecasting (ECMWF). https://climate.copernicus.eu/global-climate-highlights-2023.

The Copernicus Climate Change Service (C3S) supports the EU's efforts toward a climate-smart society by providing consistent, accurate, and validated information about the past, present, and future climate in Europe and in the rest of the world.[131]

C3S delivers long-term climate data records to monitor major climate drivers and document the fingerprints of climate variability occurring in numerous environmental parameters (e.g., surface air temperature). It also produces climate predictions and projections for relevant variables and provides the analytics tools required to transform these data into actionable insights.

C3S have recently published a set of *Global Climate Highlights for 2023*, showing that the global average temperature in 2023 was 14.98°C, 0.17°C higher than what was recorded in 2016.[132]

131 Copernicus Climate Change Service (C3S).

132 *C3S*. 2024. Global Climate Highlights 2023. 9 January.

The European Space Agency (ESA) Climate Change Initiative (CCI) aims to realize the full potential of the long-term global EO archives that ESA and its member states have established over the past 40 years, as a significant and timely contribution to the ECV databases required by UNFCCC.[133]

CCI activities include the periodic processing of the EO datasets applying the most up-to-date algorithms, together with development of improved algorithms for the ECV production from new compatible data sources.

The CCI program comprises 27 parallel projects, each focused on specific ECV data production. There is also a dedicated climate modeling project for user assessment of products, an open-data portal providing harmonized access to all products, a toolbox to facilitate combination and analysis of products, and a visualization tool raising awareness.

Map 33: Regional Differences in Sea-Level Height as Measured by Satellite Altimetry

Source: Image from the European Space Agency. https://climate.esa.int/en/projects/sea-level/.

Sea level does not rise uniformly everywhere around the planet. Measurements by satellite show how sea levels vary across the globe due to winds, atmospheric pressure, the ocean floor, the Earth's rotation, as well as the water's temperature and salinity. Such globally consistent measurements are possible only with satellite technology.

In addition to these large international EO-based climate variability initiatives, there are a few smaller, dedicated initiatives targeting the use of EO-based information to tackle climate resilience in a development context. These include the ESA EO for Sustainable Development Climate Resilience activity, which was recently completed.[134] It laid the foundations for the new ESA Global Development Assistance Climate Resilience activity, which is ongoing.[135]

EO can provide environmental information to better assess climate hazards, exposure, and vulnerability, allowing climate risks and their potential impact to be estimated, thereby better defining what actions can be taken to improve climate resilience. Hazards include floods, droughts, storms, extreme weather events, increased coastal erosion, heat waves, outbreaks and intensification of pests and diseases, increase in temperature, rising of sea level, and changes in rain patterns.

133 ESA Climate Office.

134 Earth Observation for Sustainable Development (EO4SD).

135 ESA GDA. Climate Resilience.

Figure 23: The Intergovernmental Panel on Climate Change Risk Framework

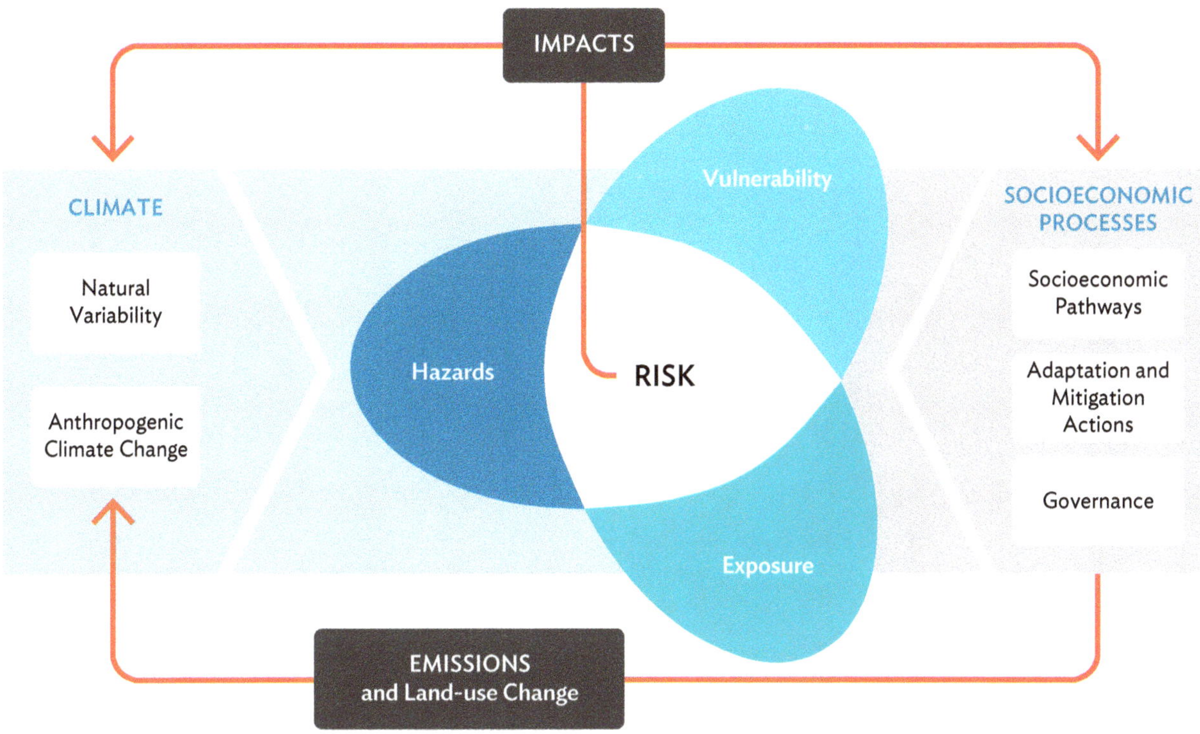

Source: IPCC.

The IPCC risk framework is a methodology that characterizes climate-related risks and their impact through the interaction of climate-related hazards with the vulnerability and exposure of human and natural systems.

Changes in both the climate system (left) and socioeconomic processes, including adaptation and mitigation (right), are drivers of hazards, exposure, and vulnerability.

The combination of climate indicators (comprising EO-based data, in-situ data, and climate projections) with socioeconomic indicators provides information about actual climate risks.

- The main EO information content consists of the variability of a wide range of environmental information for **Land** (e.g., soil moisture, snow extent, glacier extent, lake level, lake water quality, land cover [including open water body], land use, land surface temperature, evapotranspiration, drought, fire); **Ocean** (e.g., coastal sea level, water quality, met-ocean conditions); and **Atmosphere** (e.g., precipitation, air temperature, greenhouse gas concentration, air quality). Some examples are given below.

Map 34: Climate Proofing and Watershed Management for Road Infrastructure in the Manatuto-Natarbora Road Corridor in Timor-Leste

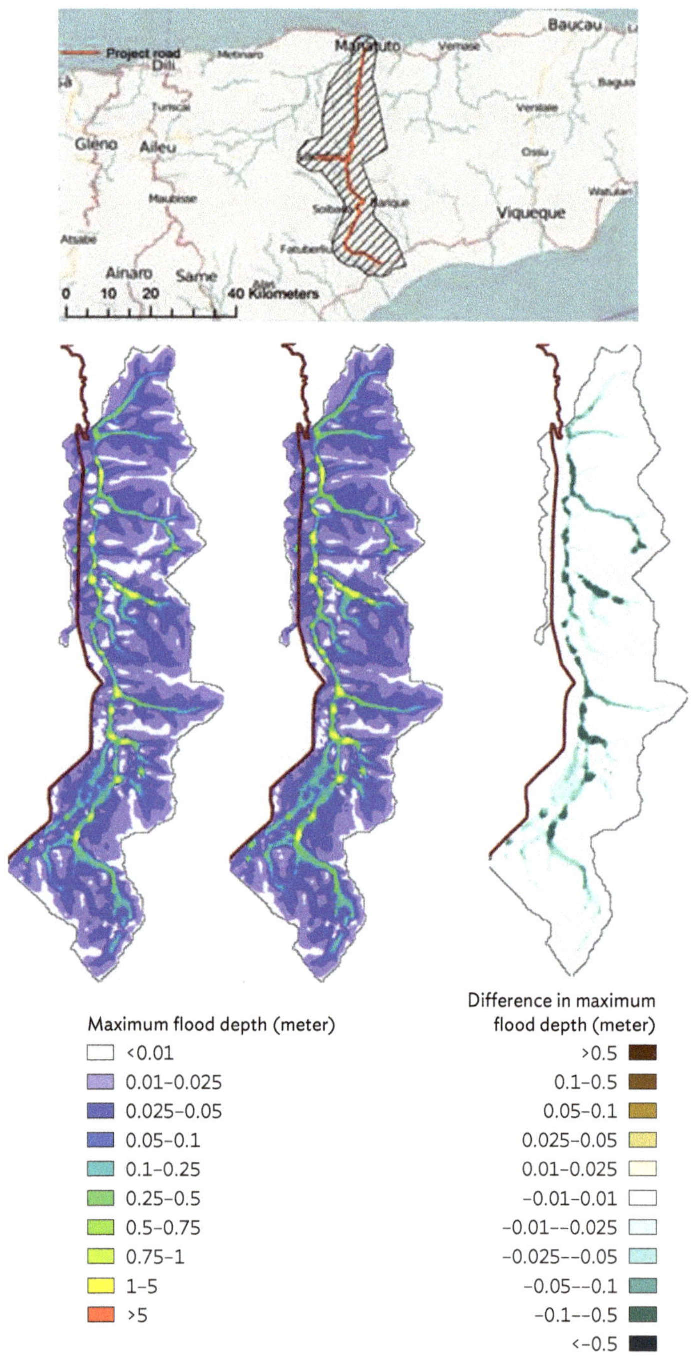

Source: Image from DHI GRAS, Denmark; processed for the European Space Agency and Asian Development Bank, reference: ADB. 2017. *Earth Observation for a Transforming Asia and Pacific*. p. 18.

(Top) Location of the Manatuto–Natarbora 81-kilometer (km) long road and its 10 km spur extending to Laclubar.

(left) Hydrodynamic modeling, showing the maximum water depth at present conditions; (middle) the projected increase in water depths for a future climate variability scenario (RCP 8.5 for 2090); and (right) the difference between the two. These techniques provide a valuable map of predicted changes to support road infrastructure future planning and maintenance.

Map 35: Example of Maximum Overland Water Flow Simulated for a 20-Year Event in 2090 Under RCP8.5 for the Southern Region of the Manatuto–Natarbora Road Corridor, Timor-Leste

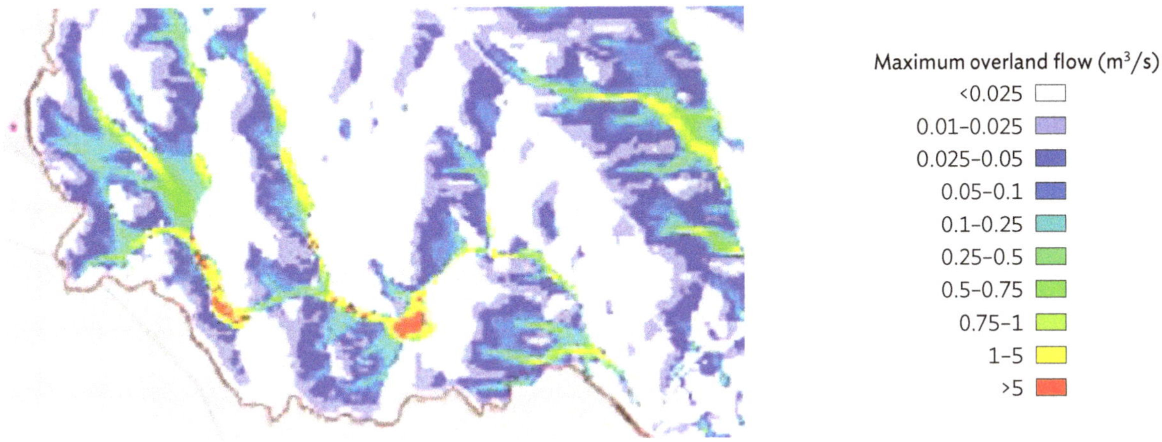

m^3/s = cubic meters per second.

Note: The water flow indicates that road sections are more susceptible to scouring and erosion due to lateral flows.

Source: Image from DHI GRAS, Denmark, processed for the European Space Agency and Asian Development Bank, reference: ADB. 2017. *Earth Observation for a Transforming Asia and Pacific.* p. 18.

Index risk maps can quickly and efficiently communicate the risk level and risk distribution across a country or region. They can provide support for preliminary risk analysis and inform the prioritization of detailed investigation and investments (image processed by DHI GRAS, Denmark).

Map 36 shows the flood hazard mapping for the coastal area around Uaitame and Irabin Leteira via a flood hazard index, which is derived through analysis of the digital topography for the areas.

The index ranks the terrain model from 1 to 6, with the highest value corresponding to areas lying lower along streams, rivers, and coastal zones, and therefore more susceptible to flooding.

Map 36: Indexing of Flood-Related Risks in Timor-Leste

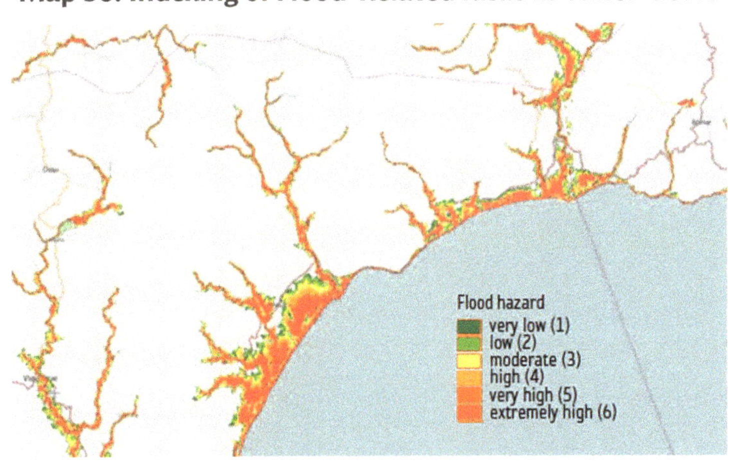

Source: Image from DHI GRAS, Denmark; processed for the European Space Agency and Asian Development Bank, reference ADB. 2017. *Earth Observation for a Transforming Asia and Pacific*. p. 19.

Map 37: Indexing of Flood-Related Risks in Timor-Leste (Population Exposure)

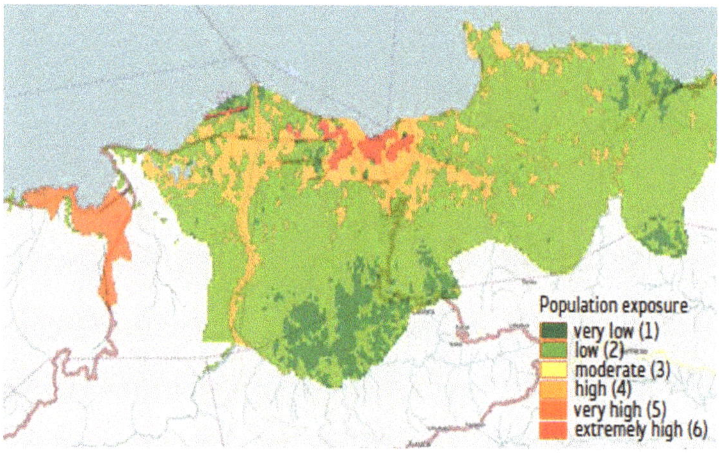

Source: Image from DHI GRAS, Denmark; processed for the European Space Agency and Asian Development Bank, reference: ADB. 2017. *Earth Observation for a Transforming Asia and Pacific*. p. 19.

Map 37 shows the population exposure index in the Dili area. A population and property index is derived from the WorldPop gridded population dataset.

The WorldPop dataset covers built-up areas and infrastructure through high-resolution EO satellite imagery (e.g., Landsat) and uses this imagery to provide more detail to the lower-level (non-space), district-level census data by redistributing and remapping this data on to a spatial grid compatible with the imagery.

The exposure index is calibrated with increasing values reflecting increasing population density.

Map 38 shows the flood risk indexing for the population in the Dili area, derived from the flood hazard and population exposure indexes previously described, at 30-meter spatial resolution.

Map 38: Indexing of Flood-Related Risks in Timor-Leste (Flood Risk)

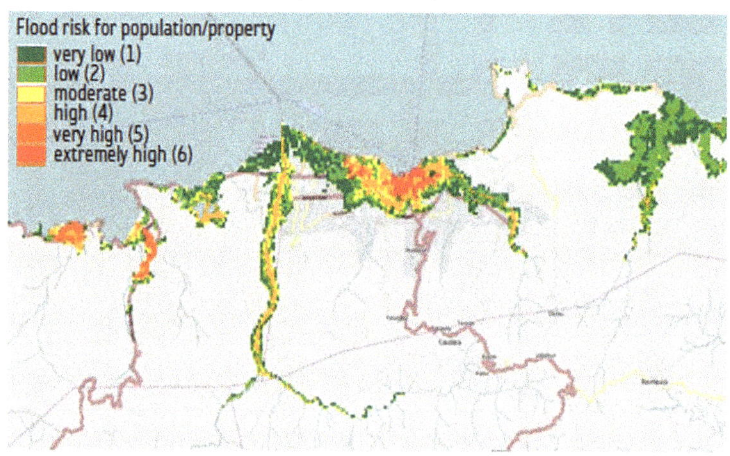

Source: Image from DHI GRAS, Denmark; processed for the European Space Agency and Asian Development Bank, reference: ADB. 2017. *Earth Observation for a Transforming Asia and Pacific*. p. 19.

Map 39: Indexing of Flood-Related Risks in Timor-Leste (Flood Risk for Roads)

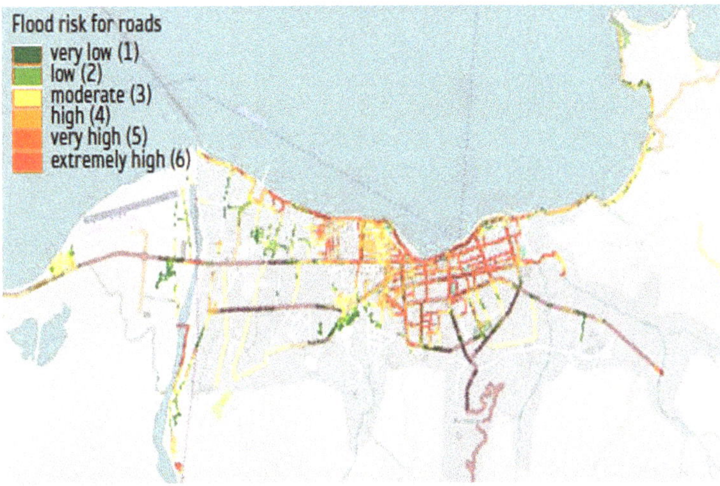

Source: Image from DHI GRAS, Denmark; processed for the European Space Agency and Asian Development Bank, reference: ADB. 2017. *Earth Observation for a Transforming Asia and Pacific*. p. 19.

Map 39 shows the flood risk indexing for roads in the Dili area. These flood risk maps provide a nationwide view of which roads are more susceptible to scouring or erosion and overtopping due to flood events.

Map 40: Example of Cyclone Track Mapping for Tropical Cyclone LOLA-23 in Vanuatu, Solomon Islands on 21–26 October 2023

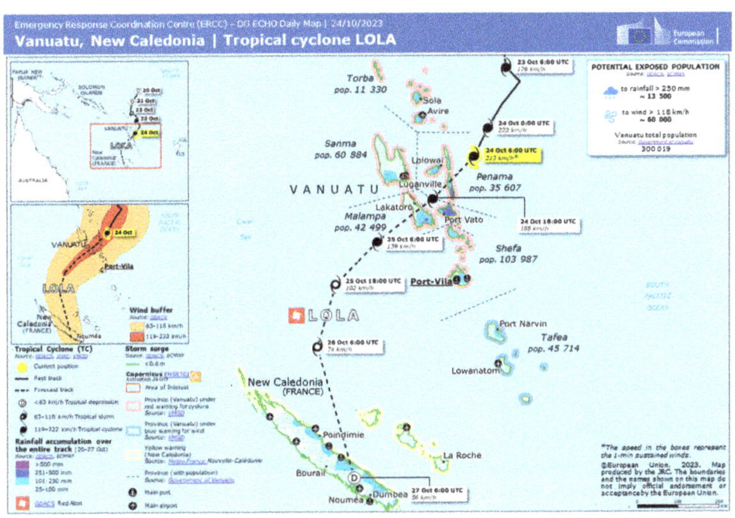

Source: Image from the Global Disaster and Alert Coordination System (GDACS), Charter activation 845 of 23 October 2023 at https://disasterscharter.org/activations/storm-hurricane-urban-in-vanuatu-activation-845-.

This map was produced by the Global Disaster and Alert Coordination system (GDACS).[136] Maximum wind topped 222 kilometers per hour (kph) (Category 4), and maximum storm surge was at 0.4 meter.[137]

The cyclone made landfall on 24 October 2023, and the Vanuatu Meteorology Department upgraded it to category 5; winds brought by Lola were estimated as having speeds of up to 230 kph at the center of the storm.

136 The International Charter: Space and Major Disasters. Tropical Cyclone Lola in Vanuatu.

137 Global Disaster Alert and Coordination System. Overall Red Alert Tropical Cyclone for LOLA-23.

Map 41: Tropical Cyclone LOLA-23, Preliminary Satellite-Derived Agricultural Assessment, Anuta Island, Temotu Vatu, Temotu Province, Solomon Islands, 7 November 2023

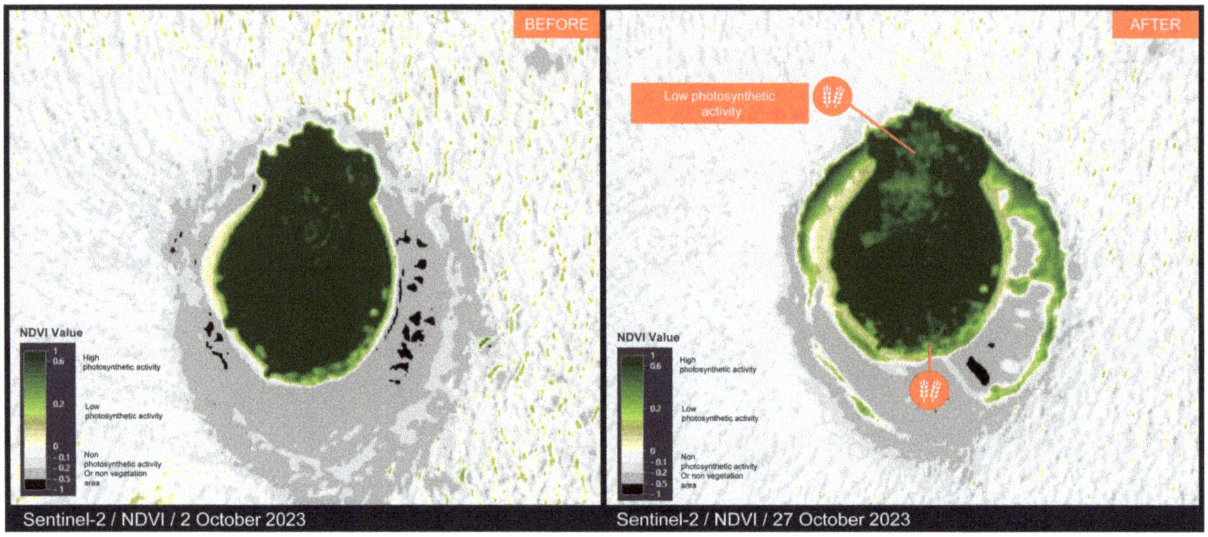

Source: Image from UNOSAT and UNITAR; contains modified Copernicus Sentinel Data 2023 from the European Space Agency. https://unosat.org/products/3725.

Map 41 is derived from Sentinel-2 imagery at 10-meter resolution before and after the cyclone passage (2 October 2023 and 27 October 2023) and produced on 7 November 2023.[138] The images show a reduction of the photosynthetic activity observed in the northern and southern parts of Anuta Island on 27 October 2023. This may indicate a potentially affected vegetation area due to the passage of the cyclone.

In summary, how are these types of climate variability related environmental information products generated from EO satellite data relevant to the development sector?

Climate variability threatens to reverse development gains, deteriorate livelihoods, trigger displacement, and worsen poverty. Sustainable development and poverty reduction cannot be achieved without climate action. Asia and the Pacific is at the center of this challenge in terms of both the climate variability's impact on people's lives and the regional potential to combat climate variability.

Climate resilience has emerged as an important framework for development policy and program formulation. Climate resilience means that people, communities, businesses, and other organizations can deal with climate variability as well as begin adapting to future climate variability. This approach has the potential to preserve development gains, and minimize environmental, social, and economic damages. Climate risks cannot be eliminated, but negative impacts on people and economies can be reduced or managed.

[138] UNOSAT. Tropical Cyclone Lola over Vanuatu and Solomon Islands.

Climate services integrate data from national and international databases, as well as maps, risk and vulnerability analyses, assessments, and long-term projections and scenarios into user-specific products that aid decision-making. The Climate Services Partnership is a platform for knowledge sharing and international collaboration aimed at promoting resilience and advancing climate service capabilities worldwide.[139] To improve in addressing climate resilience, access to high-quality information on environmental hazards is needed. EO-based environmental information as presented in this section, when combined with societal information, can provide key knowledge about climate risks and their potential impact, thereby better defining what actions can be taken to improve climate resilience.

In 2021, ADB increased its ambition for delivering climate financing up to $100 billion between 2019 and 2030, of which $66 billion will be for climate mitigation finance and $34 billion for climate adaptation finance.[140]

ADB financing has included support for digital infrastructure, digital technology industries, startups, digital services, digital technology policy and strategy, data analytics and insights, and capacity development and upskilling. For instance, ADB and the Government of the Republic of Korea agreed in May 2023 to work toward establishing an ADB–Korea Climate Technology Hub (K-Hub) in Seoul to connect ADB's developing member countries to leading-edge climate technology, experts, service providers, and other stakeholders in the climate information and technology ecosystem.[141]

In line with ADB's *Climate Change Action Plan 2023–2030*, published in November 2023,[142] EO-based environmental information can support a forward-looking, innovative, and evidence-based approach to harnessing practical experience, data, climate-modeling tools, and spatial information and technology.

Climate variability resilience and proofing are directly connected to two of the seven ADB Strategy 2030 operational priorities: C. *Tackling Climate Change, Building Climate and Disaster Resilience, and Enhancing Environmental Sustainability*, and D. *Making Cities More Livable*.

3.4 Earth Observation Use Case 4: Coastal Zone Management

The main EO information content consists of **environmental** parameters (chlorophyll concentration, transparency, sediment concentration, algal bloom occurrence); **met-ocean** parameters (sea surface temperature, surface wave parameters, coastal currents, sea level variations such as storm surges); and **hydrographic information** such as coastal bathymetry, coastline location, and changes in coastal morphology (e.g., erosion, deposition) over time. Satellite imagery also supports assessment of **coastal habitats** such as sea grass beds, coastal mudflats, mangroves, and coral reefs. Satellite monitoring can detect pollution events to support timely queueing of response assets and the compilation of pollution statistics.

---

139 Climate Europe. Climate Services Partnership.

140 ADB. 2023. 6th UN Special Thematic Session on Water and Disasters—Woochong Um. Speech. 21 March.

141 ADB. 2023. ADB, Republic of Korea Sign Agreements to Establish a Climate Tech Hub in Seoul, Expand Knowledge Exchange, and Promote PPPs. News release. 3 May.

142 ADB. 2023. *Climate Change Action Plan, 2023–2030.*

- **Coastal environmental parameters** (including Chlorophyll-A concentration in µg, transparency or turbidity if required in m-1, sediment concentration in µg/l, algal bloom detection and delineation).

Map 42: Chlorophyll Maps for the Months of February and March over the Period 2018–2020 for the Region of Lampung, Indonesia

(micrograms per m^3)

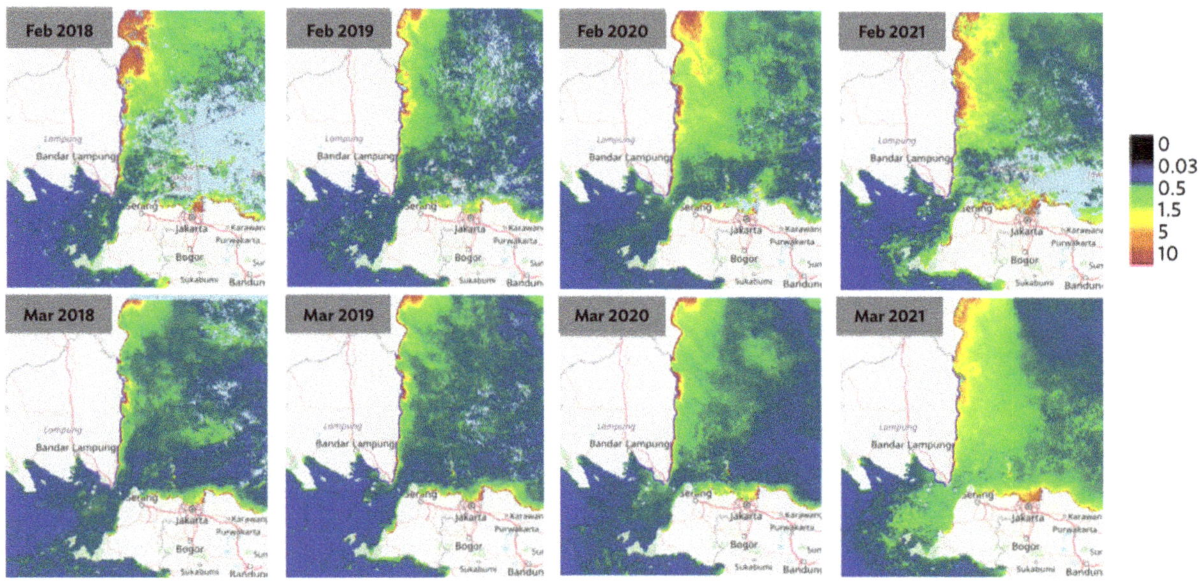

Source: Image from Planetek, for the Asian Development Bank, reference: ADB. 2023. *Earth Observation Services and Tools for Development: Examples from Indonesia*. p. 19.

This time series shows that the two regions always experience high chlorophyll concentrations in February, which decreases in March; during 2020–2021, this recurrence becomes prominent (processed from Copernicus Sentinel-3 imagery).[143]

[143] ADB. 2023. *Earth Observation Services and Tools for Development: Examples from Indonesia.*

The analysis is based on MODIS Aqua data and processed using the Ifremer Turbidity algorithm.

The high sediment concentrations (turbidity) observed may signify poor water quality and pollution, which has a negative impact on coral health and resistance.

Map 43: Yearly Mean Average in Water Turbidity in Tongatapu Island, Tonga, 2011
(in Nephelometric Turbidity Units)

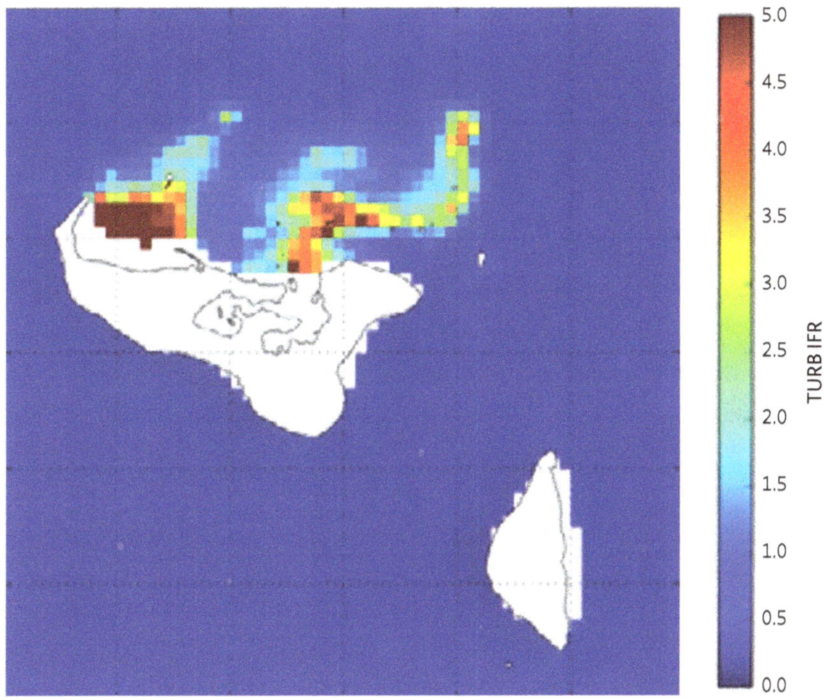

Source: Image from ACR, ITC, for the European Space Agency and Asian Development Bank, reference: ADB. 2017. *Earth Observation for a Transforming Asia and Pacific.* p. 14.

An important and substantial initiative for water quality monitoring and forecasting is that of AquaWatch Australia, currently being developed by Australia's National Science Agency (CSIRO), together with national partners from industry, research, and government.[144]

The AquaWatch data system integrates both in-situ and satellite water quality sensor technologies and uses advanced data analytics through cloud computing and AI to produce water quality monitoring and forecast information tailored to the needs of end users. The early warning forecasting combines hydrodynamic and biogeochemical modeling tools to provide accurate forecasts of water quality in diverse aquatic systems including rivers, lakes, and coastal waters. This includes harmful events such as toxic algal blooms; blackwater; offshore harmful algae blooms; and accidental sewage, farm, and mine site discharges.

[144] Commonwealth Scientific and Industrial Research Organisation (CSIRO). AquaWatch Australia.

Map 44: Example Map of Water Quality Modeling from AquaWatch Australia, July 2024

Source: Image from Commonwealth Scientific and Industrial Research Organisation (CSIRO). https://www.csiro.au/en/about/challenges-missions/AquaWatch.

A key element of the AquaWatch system is the real-time measurement of numerous water quality parameters (such as temperature, pH, dissolved oxygen, turbidity, salinity, and nutrient levels) through direct deployment of Hydraspectra optical in-situ sensors into water bodies. Hydraspectra sensors measure water reflectance and surface photos at 15-minute intervals, offering continuous monitoring capabilities to facilitate tracking of algal bloom or sediment plume evolution.

This precise local measurement data is combined with larger coverage (but less frequent) EO satellite data. The EO data being used includes Copernicus Sentinel-3 as well as CyanoSense, the first Australian-manufactured hyperspectral sensor launched on a Skykraft satellite in June 2023. Cyanosense has been developed by CSIRO specifically to differentiate harmful cyanobacteria from other algae in coastal and inland waterways.

Early evaluation of the AquaWatch data system is being carried out through a number of pilot test sites in Australia situated at Lake Hume, on the New South Wales and Victorian border, Lake Tuggeranong in the Australian Capital Territory, and the Grahamstown Reservoir.

The Government of Australia and Government of Indonesia announced an ambitious plan for water resource projects under a memorandum of understanding signed by both countries in 2023, with lake management being the main focus. As part of this, a proposed AquaWatch pilot at Indonesia's Lake Tempe is under discussion.

In late 2024, CSIRO launched a project in Viet Nam to demonstrate AquaWatch Australia for a test site at An Lao, Hai Phong.[145] The initiative aims to empower local shrimp farmers with near-real-time data and forecasts to protect shrimp stocks by helping them to anticipate challenges such as algal blooms and nutrient imbalances, thereby improving local water management practices.

The Viet Nam test site joins other global demonstration sites for AquaWatch located in Australia, Chile, Malaysia, Italy, the UK, and the US.

- **Coastal met-ocean information** such as sea surface temperature in degrees Kelvin (K), coastal wave information, and coastal currents, including the delineation of boundaries between different coastal water bodies, and the boundary between turbid coastal waters and (the more transparent) open sea, as well as frontal structures linked to river outflow, upwelling/downwelling and coastal currents. The maps below were processed from Copernicus Sentinel-3 data.[146]

Map 45: Evolution of Coastal Sea Surface Temperature in the Benguela Upwelling System, Poyeaymai Coastal Region, Cambodia

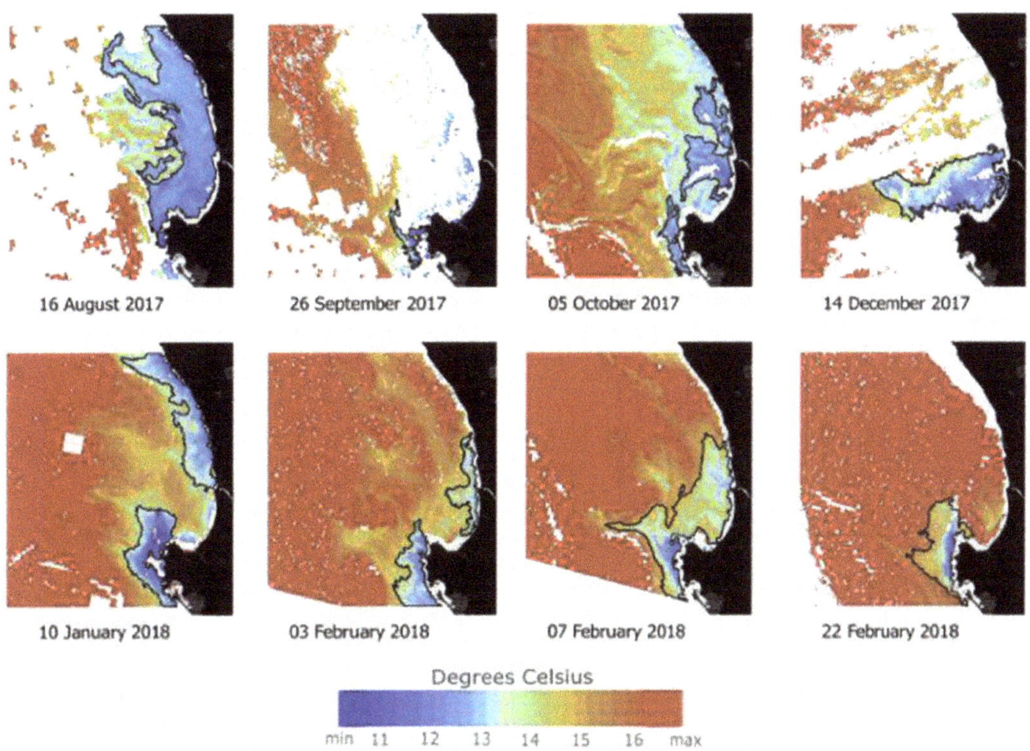

Note: Maps produced using Copernicus Sentinal-3 data.

Source: Images from K. Karamvasis et al. 2018. A methodology for monitoring the upwelling phenomenon using Sentinel-3 products. Paper prepared for the Sixth International Conference on Remote Sensing and Geoinformation of the Environment (RSCy2018). 6 August.

145 *CSIRO*. 2024. Vietnam partners with Australia to pioneer water quality monitoring. 17 December.

146 K. Karamvasis et al. 2018. A methodology for monitoring the upwelling phenomenon using Sentinel-3 products. Paper prepared for the Sixth International Conference on Remote Sensing and Geoinformation of the Environment (RSCy2018). 6 August.

Map 46: Spatial Distribution of Wave Directions (Scaled by Significant Wave Height) Along the Coastline of Georgia from the Black Sea

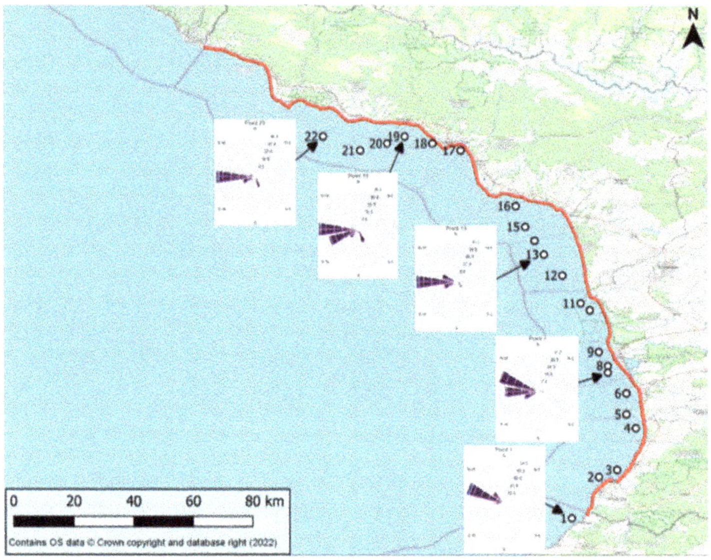

Source: Image from the European Space Agency. https://gda.esa.int/wp-content/uploads/2023/07/ESA_GDA_CR_Brochure.pdf, p. 18.

This map was produced by the Monitoring Forecasting Centre (BS MFC), which provides regular and systematic information about sea conditions for the Black Sea.

The BS MFC system is distributed by the Copernicus Marine Environment Monitoring Service (CMEMS).[147] It is based on a numerical ocean model assimilating in-situ and satellite data.

The predominant wave direction is from the west, with the largest wave heights occurring between October and February.

Map 47: Distribution of Extreme Significant Wave Heights Along the Coastline of Georgia

(in meter)

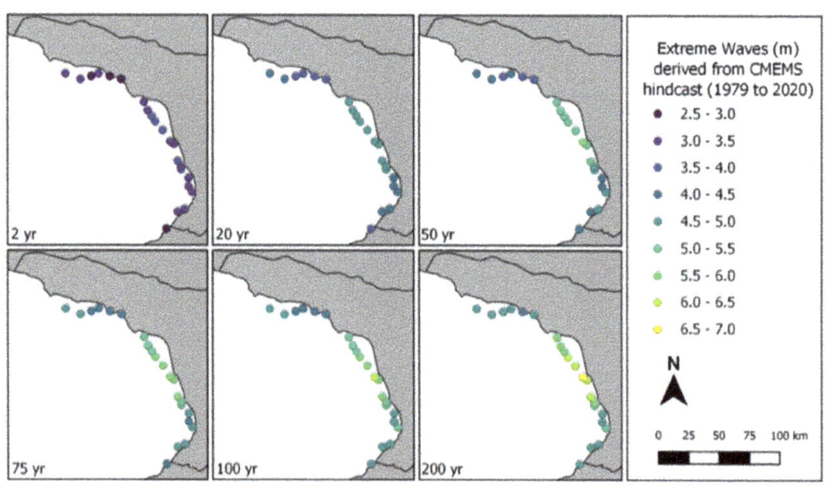

Source: Image from the European Space Agency. 2023. *Global Development Assistance: Climate Resilience.* p. 18.

These maps were derived from CMEMS hindcast data.

[147] Copernicus. Marine.

- **Coastal hydrographic information** including bathymetry and the evolution of coastline mapping to monitor erosion and deposition dynamics.

Map 48: Coastal Bathymetry Mapping for a River Outlet near Hoi An, Viet Nam, 2013

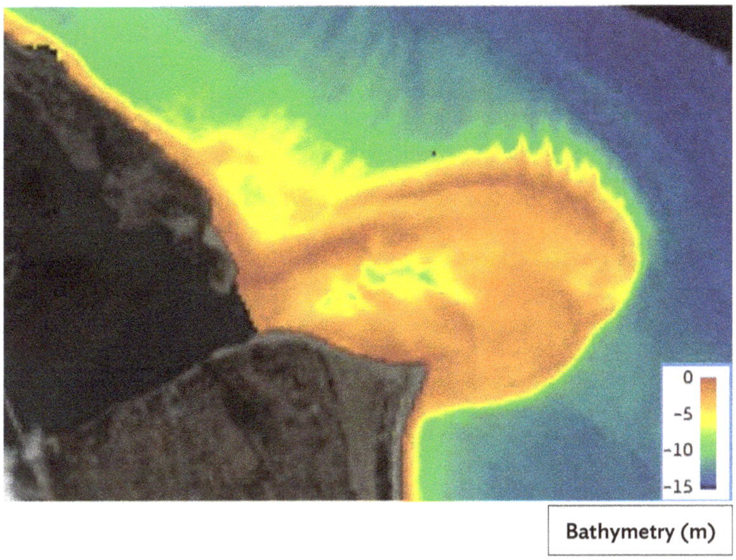

Source: Image from the European Space Agency, reference: ADB. 2017. *Earth Observation for a Transforming Asia and Pacific.* p. 46.

This map is based on medium-resolution data (Landsat-8, 30 m resolution) (processed by Geoville, Austria and DHI, Denmark; image from Landsat-8 data © USGS).

Map 49: Shallow Areas and Coastline Changes in Hoi An, Viet Nam, 2006–2013

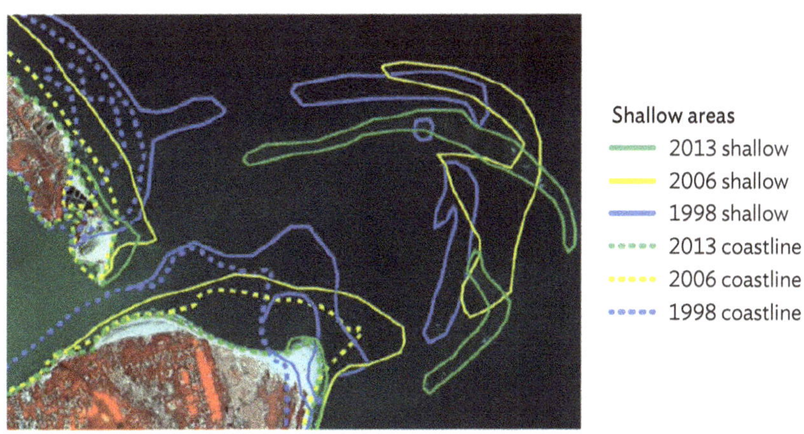

Source: Image processed by Geoville, Austria and DHI, Denmark; Image from SPOT 2, SPOT 5, SPOT 6 © CNES, distribution Airbus DS, reference: ADB. 2017. *Earth Observation for a Transforming Asia and Pacific.* p. 46.

This map presents the dynamics of sediment transport, and the areas active in this process. This is important in assessing coastline changes and in supporting vessel navigation in the area.

Map 50: Coastal Bathymetry Mapping Along the East Coast of Chu Lai Airport and Around Dung Quat Port, Viet Nam

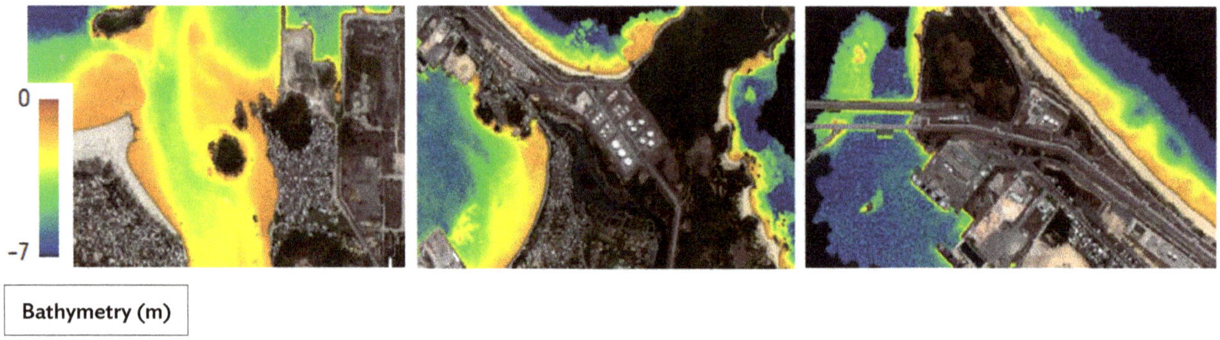

Source: Images from the European Space Agency, processed by Goeville, Austria and DHI, Denmark; (image from WorldView-2 data © DigitalGlobe), reference: ADB. 2017. *Earth Observation for a Transforming Asia and Pacific*. p. 46.

Maps are based on high-resolution data (WorldView-2, 2 m resolution) along the east coast of Chu Lai Airport and around Dung Quat Port, Viet Nam.

Map 51: Distribution of Net Areas of Erosion (in red) and Accretion (green), along the Black Sea Coastline of Georgia

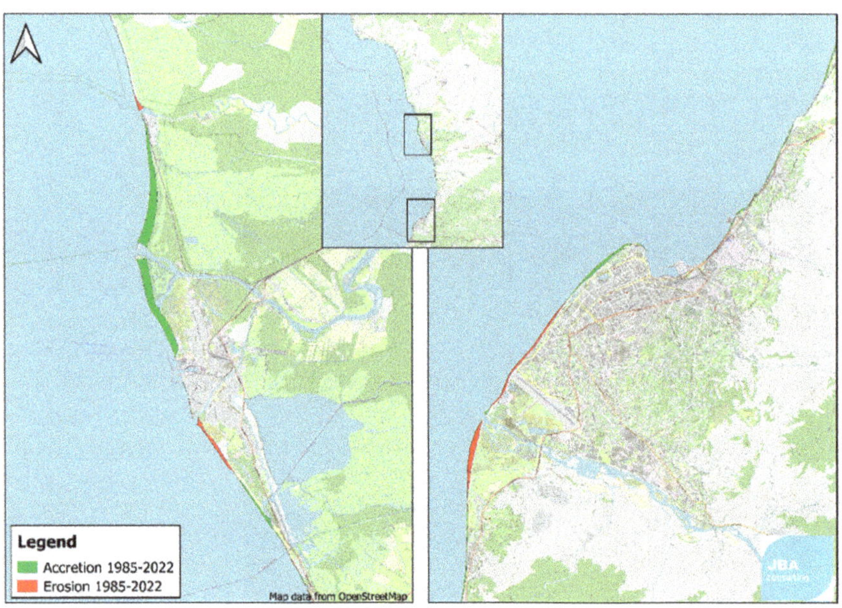

Source: Image from European Space Agency. 2023. *Global Development Assistance: Climate Resilience*. p. 18.

Shoreline positions for the last 40 years at a temporal frequency of 3 years have been mapped using the annual median composites of Landsat 5, 7, 8 and Sentinel-2 along the entire coastline.

To the immediate north of Poti, the Rioni Delta is accreting, which is a typical response to a natural sediment supply from the Rioni River.

Map 52: Coastline Variation Trend in the Region of Nadogoloa, Fiji, 2017–2022
(m/year)

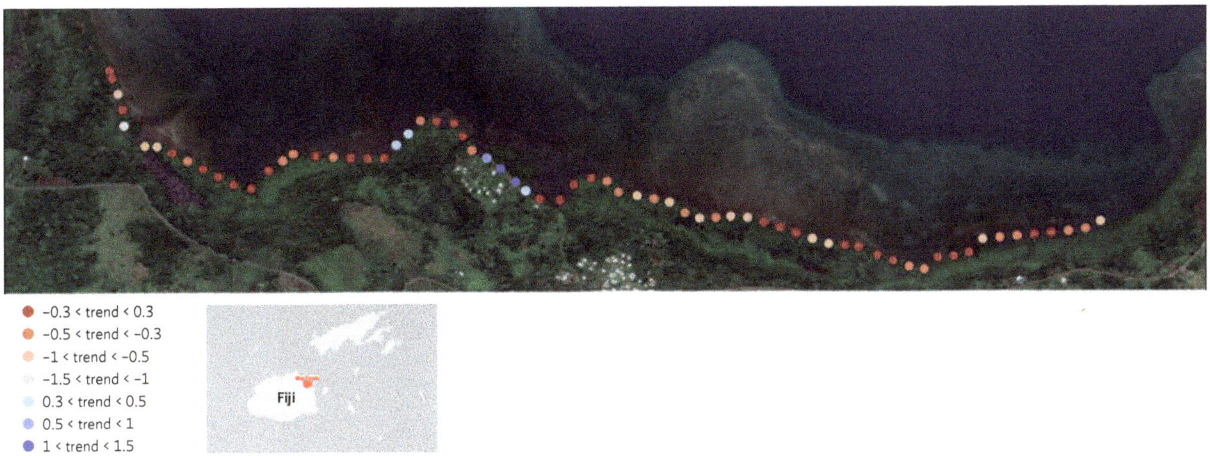

Source: Image from the European Space Agency, the shoreline variation was processed from Copernicus Sentinel-1 and Sentinel-2 time-series imagery. https://www.eo4sd-forest.info/.

The Batumi spit is accreting, while the areas around the Chorokhi River are eroding, possibly in response to a disruption in the sediment supply from hydropower infrastructure in the Chorokhi River system. The shoreline variation was processed from Copernicus Sentinel-1 and Sentinel-2 time-series imagery.[148]

- **Coastal habitat status** including mangroves, coastal mud flats, sea grass beds, and coral reefs. The information layers can include the spatial extent and the level of fragmentation of the coastal habitat, the identification of subspecies within the habitat, the identification of coastal elements subject to degradation or improvement, and the delineation of elements subject to change with respect to a defined previous reference period (e.g., loss of habitat area, encroachment of transport infrastructure, or other aspects pertaining to human activities).

The mangrove variation was processed from Copernicus Sentinel-1 and Sentinel-2 time-series imagery. Three types of mangrove habitat change along the coastline are color coded as stable (orange), loss (red), and gain (green) (Map 53).

The Global Mangrove Watch (GMW) was established in 2011 under the Japan Aerospace Exploration Agency's Kyoto and Carbon Initiative.[149] It is an international collaboration led by Aberystwyth University (UK) and solo Earth Observation (Japan), together with Wetlands International, International Water Management Institute (Sri Lanka), and UNEP World Conservation Monitoring Centre (UK).

The GMW dataset 3.0 was released in August 2022.[150] It provides maps of estimated global extent and changes of mangrove forests derived for the year 1996 and then annually for 2007–2010 and 2015–2020. This results in a total of 11 global mangrove maps, each provided at a spatial resolution of 25 m.

148 EO4SD. Forest Management.

149 Advanced Land Observing Satellite. The K&C Global Mangrove Watch.

150 Global Mangrove Watch.

Map 53: Evolution of Coastal Mangrove Habitat Extent in the Region of Nadogoloa, Fiji, 2017–2022

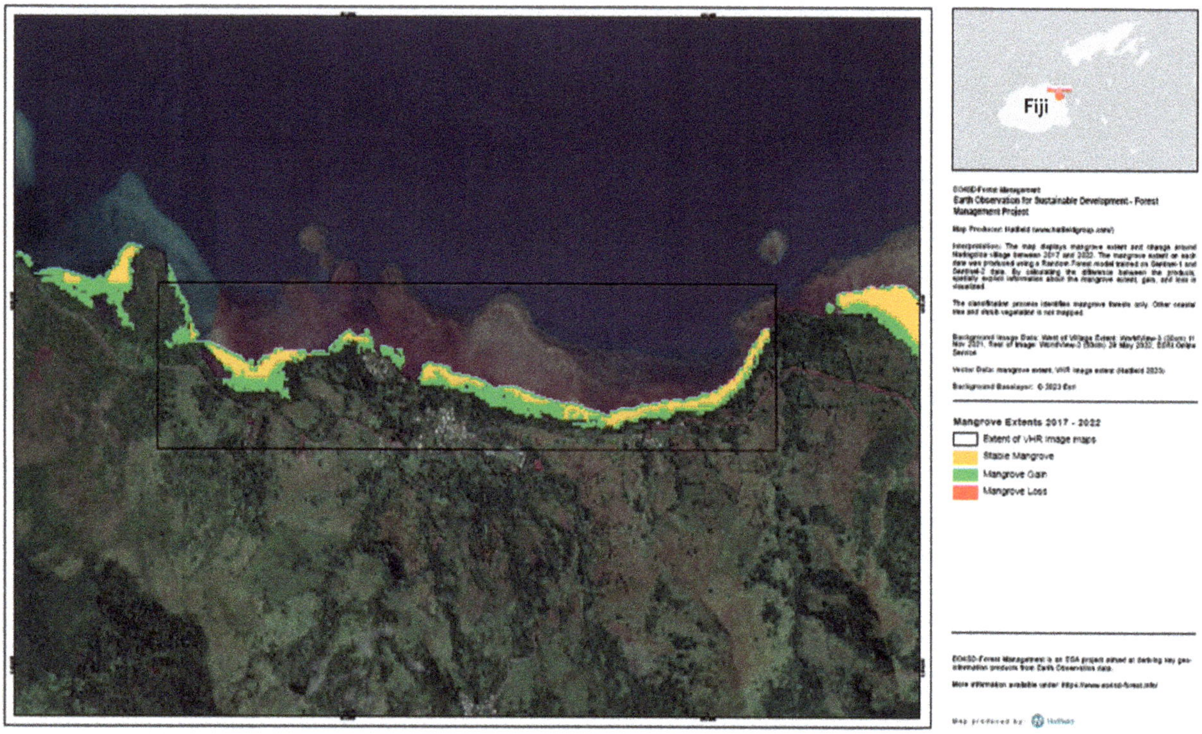

Source: Image from the European Space Agency. https://www.eo4sd-forest.info/.

Map 54: Example of Global Mangrove Watch Global Map

Source: Image by Global Mangrove Watch interactive map site version 3.0, from the Japan Aerospace Exploration Agency. https://www.globalmangrovewatch.org/.

The mangrove maps are publicly available through the World Resources Institute Global Forest Watch portal (http://www.globalforestwatch.org) and for free download via the UNEP World Conservation Monitoring Centre, Ocean Data Viewer.

Cloud cover limits the use of optical satellite data in tropical and subtropical regions. However, synthetic aperture radar (SAR) sensors are well suited for mangrove monitoring as SAR data can be acquired both day and night, and in the presence of cloud cover. L-band SAR is particularly suitable as it is sensitive to both vegetation structure and to detect changes via a time series of repeated acquisitions of the same area of ground. JAXA's JERS-1 SAR, ALOS PALSAR, and ALOS-2 PALSAR-2 missions provide a coherent and consistent L-band SAR data series spanning over 2 decades. This opens a unique opportunity to map the extent and changes of the global mangrove cover over the past 20 years.

This information is of key importance for countries reporting on the UN SDGs or participating in the Reducing Emissions from Deforestation and Forest Degradation in Developing Countries (REDD+) program developed by the United Nations Framework Convention on Climate Change (UNFCCC).

Map 55: Example of Mangrove Map Analysis

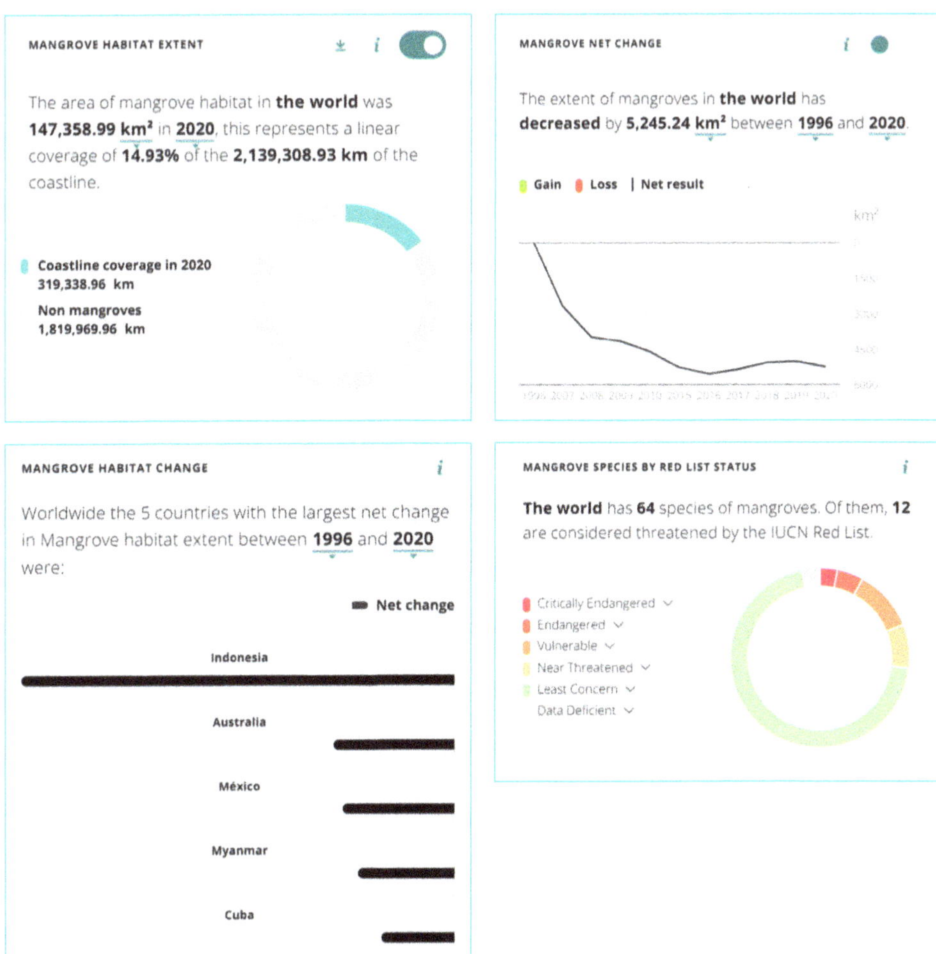

Source: GMV interactive map (version 3.0; JAXA). https://www.globalmangrovewatch.org/.

In summary, how are these types of coastal-related environmental information products generated from EO satellite data relevant to the development sector?

The world's oceans generate a wide range of goods and services and are essential to the global economy. The blue economy supports economic growth, social well-being, and improvement of livelihoods through employment and creation of jobs, and the continued health of coastal and marine ecosystems and the services they provide.

This approach requires careful planning and management based on a solid understanding of marine and coastal ecosystems that are often rich and varied, with high levels of productivity and biodiversity. These environments are also vulnerable to climatic and anthropogenic pressures, affecting resilience of the living marine resources and the services they provide.

Marine environments are expensive and challenging to measure and monitor using in-situ techniques alone. In many cases, there are no viable alternatives to satellite EO for providing the information decision-makers need to build a sustainable blue economy.

The range of coastal environmental information given as examples in this section can be used in different ways to address different coastal management aspects relevant to developing countries:

- For routine monitoring of water quality parameters to detect anomalies (e.g., algal blooms, temperature anomalies).
- As historical statistics on local environmental physical conditions and habitats to perform "nowcasting" and forecasting of coastal conditions for environmental impact assessment and for management of coastal facilities (e.g., aquaculture sites, desalination plants, water treatment facilities).
- For the analysis of longer-term trends and indicators (e.g., for the assessment of climate resilience and proofing).

Coastal and marine mapping, monitoring, and management are directly connected to two of the seven ADB Strategy 2030 operational priorities: C. *Tackling Climate Change, Building Climate and Disaster Resilience, and Enhancing Environmental Sustainability*, and E. *Promoting Rural Development and Food Security*.

3.5 Earth Observation Use Case 5: Forest Monitoring and Management

The main EO information content consists of basic forest type maps, basic forest cover change mapping, and a range of more advanced information products regarding forest biomass and carbon statistics. Both multispectral optical and radar imagery are used in generating these forest information products.

- **Basic forest type mapping.** For managed forests, typically, two to four volume classes within deciduous and coniferous forests can distinguished. For tropical forest, the classes are customized by distinguishing typically broad-leaved forest, bamboo, palm forest, mixed forest, or forest plantations.

Map 56: Forest Ecosystems in Dilijan National Park, Northern Armenia, 31 July 2019

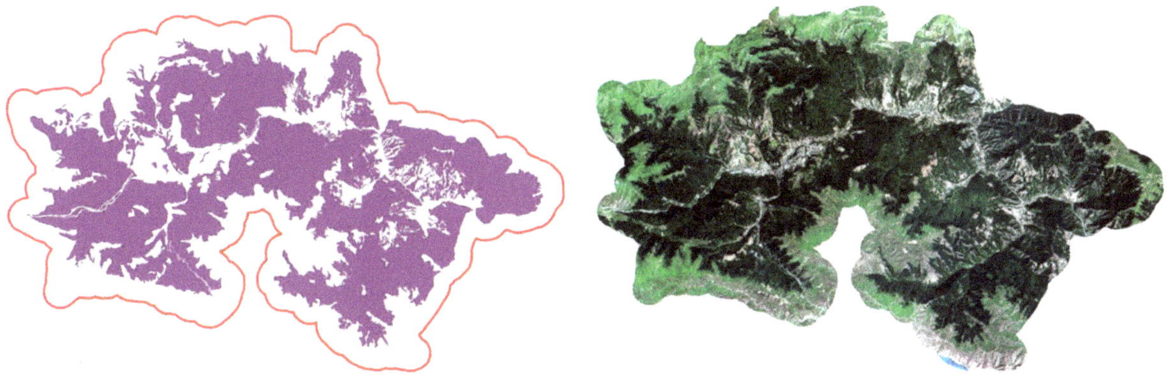

Note: Left, area of interest in red outline, national park limits in purple; right, Sentinel-2 optical image of the area of interest.

Source: Image from European Space Agency. 2020. *Characterization of Dilijan National Park Forest Ecosystems, Armenia*. 2 September. p. 12.

Map 57: Tree Cover Mask and Tree Type Map for the Dilijan Area of Interest, 2019

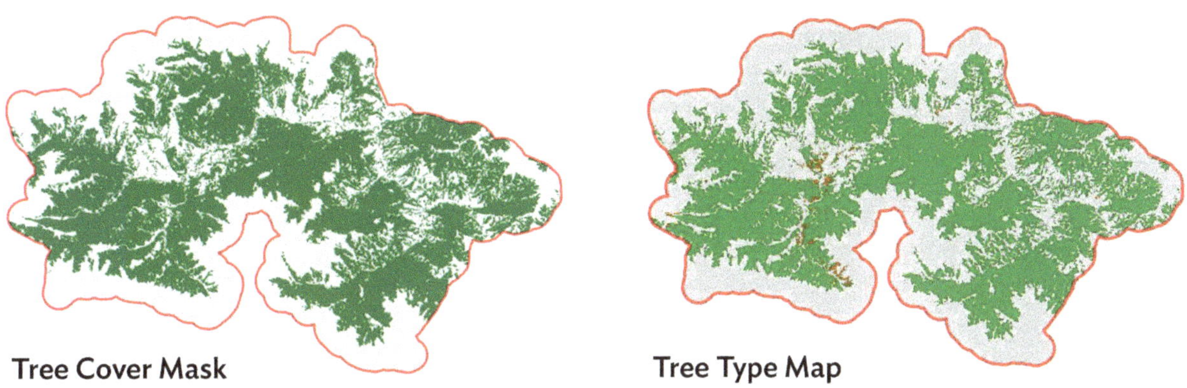

Source: Image from European Space Agency. 2020. *Characterization of Dilijan National Park Forest Ecosystems, Armenia*. 2 September. p. 26 and p. 31.

(Left) Tree cover mask for the Dilijan area of interest for 2019 at 10 m resolution with binary layer of tree/non-tree, green/grey; (Right) Forest type map for 2019 at 10 m resolution with grey: non-forest; green: broadleaved trees; brown: coniferous trees.[151]

[151] EO Science for Society. EO Clinic: Characterization of Dilijan National Park Forest Ecosystems, Armenia.

Map 58: Tree Cover Density Map for the Dilijan Area of Interest, 2005

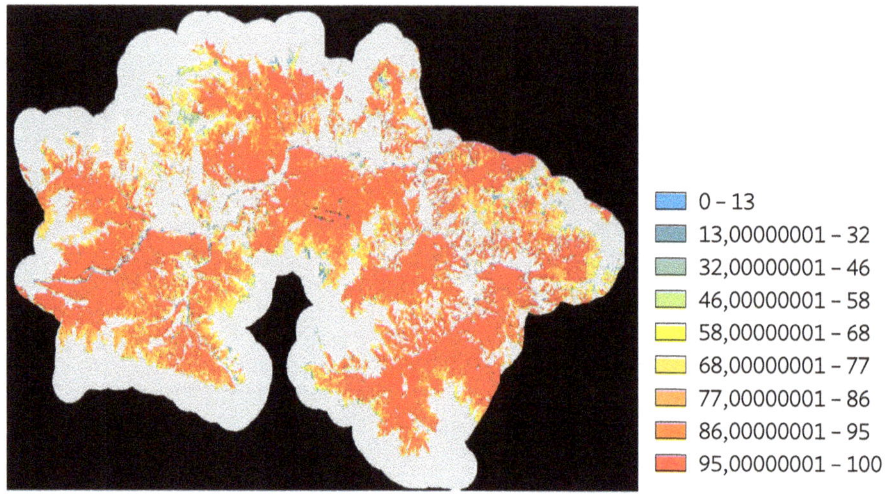

Source: Image from European Space Agency. 2020. *Characterization of Dilijan National Park Forest Ecosystems, Armenia.* 2 September. p. 29.

Tree cover density map for 2005, defined as the proportion of tree canopy coverage per pixel, in the range of 1% to 100% values.

- **Forest cover change mapping** provides information on afforestation, deforestation, and reforestation (e.g., canopy gaps, clear cuts, burnt areas, and regrowth). These changes occur typically as a result of illegal logging, selective logging, and/or storm damage).

Map 59: (Left) Tree Cover Change Mask; (Right) Deforestation and Forest Degradation Map, Area of Interest for the Dilijan National Park, Northern Armenia

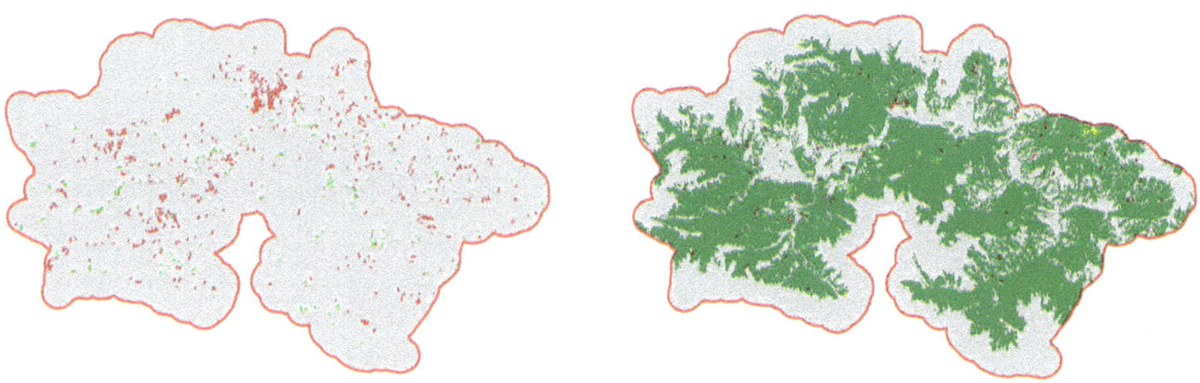

Source: Image from European Space Agency. 2020. *Characterization of Dilijan National Park Forest Ecosystems, Armenia.* 2 September. p. 27 and p. 33.

(Left) Tree cover change mask over the period 1991–2019 at 30 m resolution (red points: deforestation, green points: reforestation); (Right) Deforestation and forest degradation map over the period 2005–2010 (grey: stable non-forest; green: stable forest; red: deforestation; yellow: forest degradation).

Map 60: Forest Disturbances Map Using Sentinel-1 Radar Image Data at the Border Between the Lao People's Democratic Republic and Viet Nam

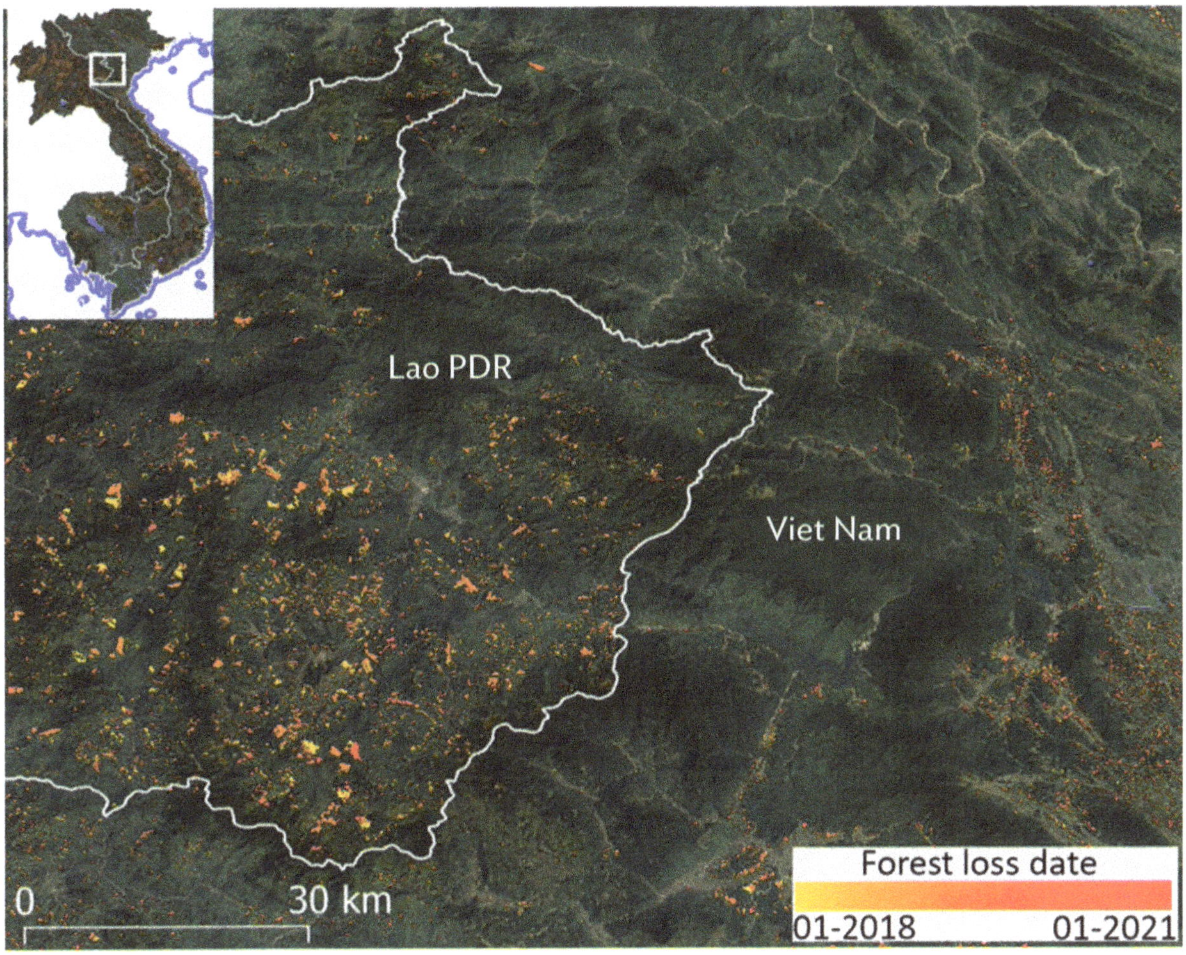

Lao PDR = Lao People's Democratic Republic.

Source: Image from S. Mermoz, T. Le Toan, and A. Bouvet. 2021. *Sentinel-1 for Observing Forests in the Tropics – SOFT*. Final Report. 22 April. European Space Agency. p. 31.

This map of forest disturbances was produced using a time series of Sentinel-1 radar image data from the end of 2017 to the beginning of 2021.[152]

The underlying image is from Google Earth. The map shows marked differences in forest changes between countries with high losses of forest currently happening in northern Lao People's Democratic Republic (Lao PDR), and low forest losses taking place in northern Viet Nam.

152 EO Science for Society. Sentinel-1 for Observing Forests in the Tropics (SOFT).

This map presents the forest loss detection results in detail, highlighting the various sizes and distributions of disturbed areas across various sites in Cambodia, the Lao PDR, and Viet Nam.

Map 61: Forest Loss Detection in Disturbed Areas Across Various Sites in Cambodia, the Lao People's Democratic Republic, and Viet Nam

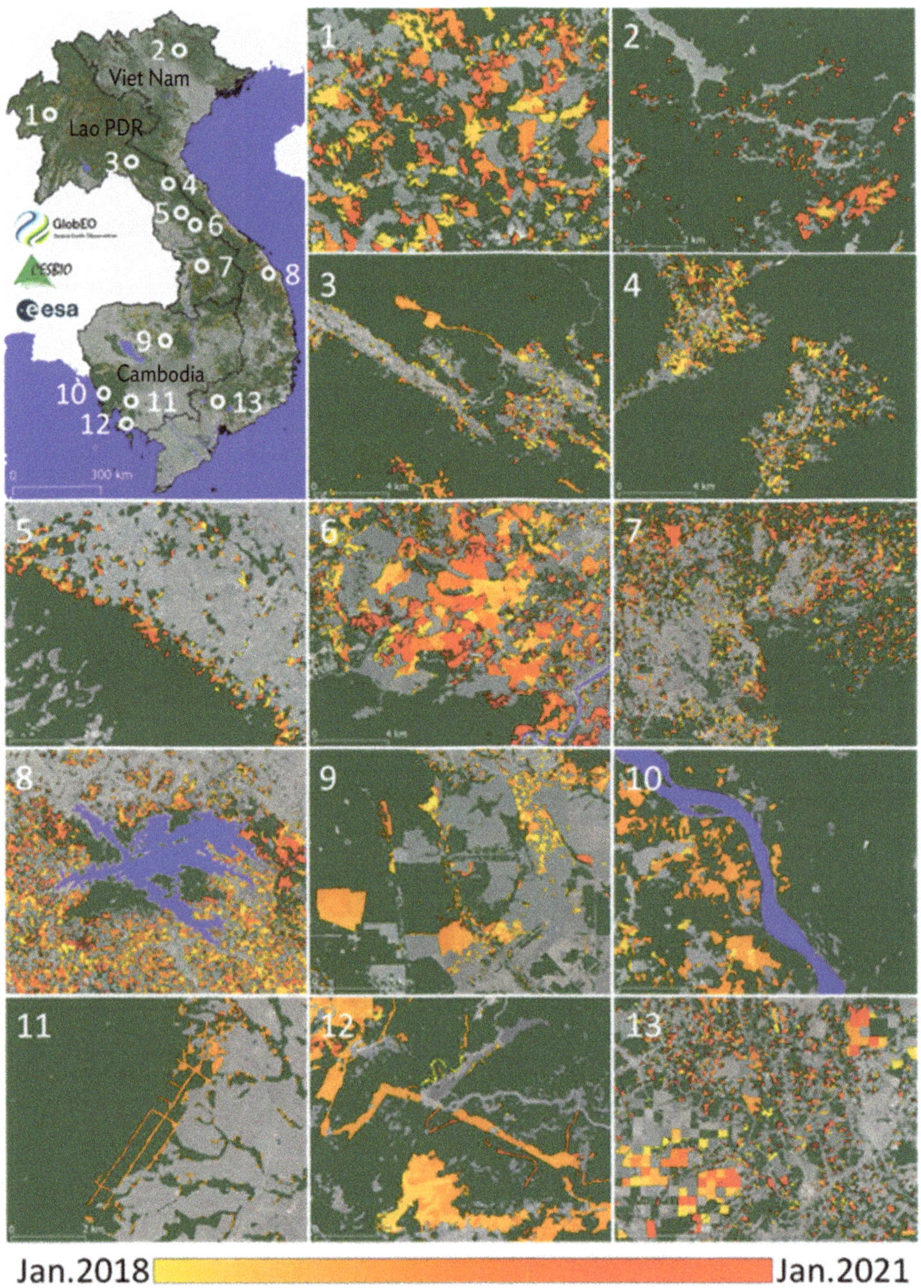

Lao PDR = Lao People's Democratic Republic.

Source: Image from S. Mermoz, T. Le Toan, and A. Bouvet. 2021. *Sentinel-1 for Observing Forests in the Tropics – SOFT.* Final Report. 22 April. European Space Agency. p. 32.

- **Forest biomass and carbon statistics.** The determination of forest biomass and carbon stock has become increasingly more important as activities of the UN Collaboration on Reduction of Emissions from Deforestation and Degradation (REDD and REDD+) increase. At present, this is mostly done from in-situ measurements on tree types and volume and height estimates. Advanced EO products can be delivered on tree stem volume in tropical as well as boreal forests with typically 30 m resolution on a large spatial scale. The EO techniques use radar to penetrate into the forest canopy where the return signal is used to derive information on stem volume. The EO-derived stem volume can then be related to aboveground biomass and carbon stock by calibrating and correlating with in-situ data.

Map 62: Forest Height Estimation for the Area Surrounding Medan and Banda Aceh in North Sumatra, Indonesia

Source: Image from Synspective, Japan. https://synspective.com/solutions/fim/.

The tree height is computed via AI using ALOS-2 radar (SAR) coherence and GEDI (Global Ecosystems Dynamics Investigation) Lidar measurements as training data (specifically, the Level 2A Geolocated Elevation and Height Metrics Product).[153]

153 Synspective. Forest Inventory Management (FIM).

The estimate is computed using ALOS-2 image intensity (backscatter coefficient) and biomass models. Forest carbon stock can then be estimated directly as 50% of these biomass figures.

Map 63: Aboveground Biomass Estimation at 10–30 Meter Resolution for the Area Surrounding Medan and Banda Aceh in North Sumatra, Indonesia (tonnes/hectare)

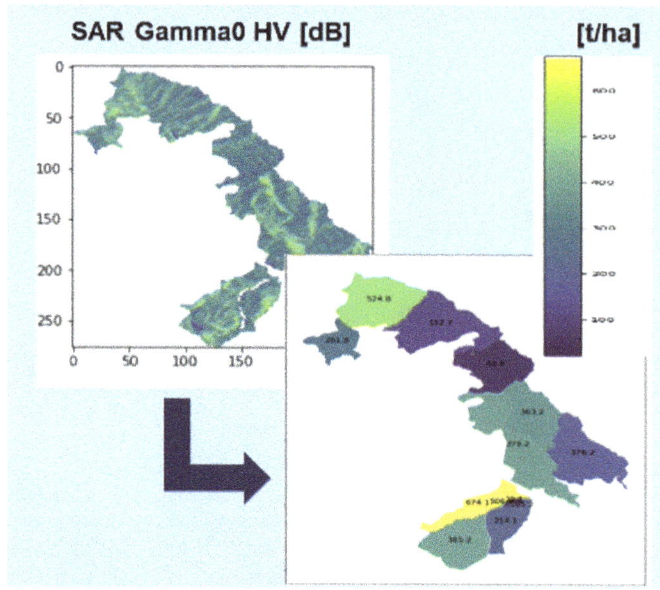

Source: Image from Synspective, Japan, reference at https://synspective.com/solutions/fim/.

Map 64: Aboveground Biomass Map over Kalimantan, Indonesia, 2008 (tonnes/hectare)

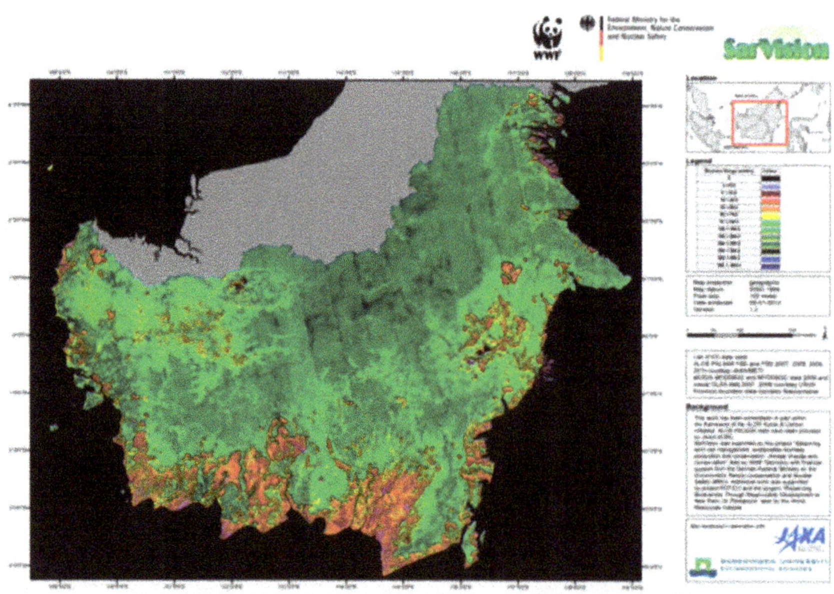

Source: Image from European Space Agency. 2013. *Earth Observation for Green Growth: An Overview of European and Canadian Industrial Capability*. p. 26.

The spatial resolution of the map information is 50 m, derived from LIDAR and radar sources, ICESat-GLAS and ALOS PALSAR.

Map 65: Global Aboveground Biomass Estimates, 2020
(tonnes/hectare)

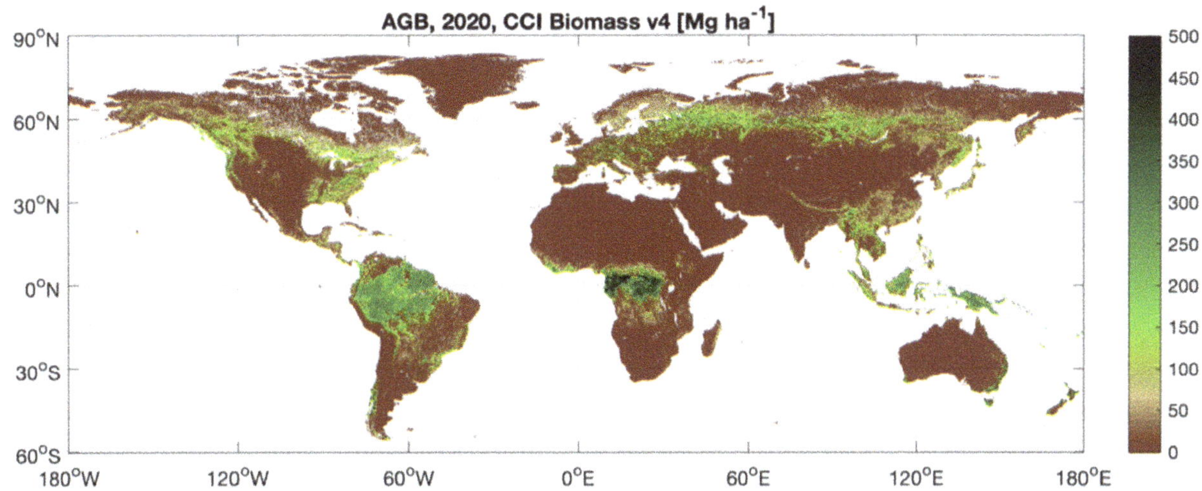

Source: ESA Climate Change Initiative. https://catalogue.ceda.ac.uk/uuid/95913ffb6467447ca72c4e9d8cf30501/.

The products are generated at 100 m spatial resolution and are based on ESA's C-band (Sentinel 1A and Sentinel 1B) and JAXA's L-band synthetic aperture radar (ALOS-2 PALSAR-2), with additional information from space borne LIDAR (e.g., NASA's Global Ecosystem Dynamics Investigation Lidar GEDI).[154]

In summary, how are these types of forest-related environmental information products generated from EO satellite data relevant to the development sector?

The global forestry sector faces many challenges, particularly in developing countries. Rapidly growing human populations require more land for agriculture, while demand for commodities such as timber, palm oil, and soy are driving both legal and illegal deforestation, degradation, habitat, and biodiversity loss, and depriving forest-dependent communities of their livelihoods. The loss of forests also leads to huge carbon emissions which are driving climate variability. The Global Climate Observing System (GCOS) considers aboveground biomass as an essential climate variable due to its functions as both a source of atmospheric CO_2 (and other greenhouse gases) when forest is lost, and as a sink for CO_2 due to forest growth.

Key development issues can be better addressed using satellite EO in the forestry sector through the following:

- Forest cover mapping and support assessment of deforestation, forest degradation, and forest carbon tracking in emission reduction programs, to address loss and depletion of forests.
- Forest resource management and support to forest stock growth estimation, to address destruction of forest habitats and loss of livelihood opportunities.
- Support to forest accreditation and certification schemes to improve economic development opportunities that result from good forest sector governance.

[154] ESA Climate Office. Data.

- Integrating forests into sustainable economic development via ecosystem service assessments and biodiversity assessments.
- Estimating forest carbon biomass to support sustainable development, putting an economic value on the environment and recognizing the importance of natural capital.

Forest mapping, monitoring, and management are directly connected to two of the seven ADB Strategy 2030 operational priorities: C. *Tackling Climate Change, Building Climate and Disaster Resilience, and Enhancing Environmental Sustainability*, and E. *Promoting Rural Development and Food Security*.

3.6 Earth Observation Use Case 6: Greenhouse Gases Mapping and Monitoring (CO_2, CH_4, N_2O)

The main EO information content consists of the global spatial location, distribution, and concentrations of the main greenhouse gas (GHG) sources and emissions—carbon dioxide (CO_2), methane (CH_4), and nitrous oxide (N_2O).

Carbon dioxide is released through natural processes, such as volcanic eruptions, plant respiration, and animal and human breathing. However, the atmospheric CO_2 concentration has increased by 50% since the Industrial Revolution began in the 1800s, due to human activities like the burning of fossil fuels and large-scale deforestation. Due to its abundance, CO_2 is the main contributor to climate variability.

Methane is produced naturally through decomposition. However, human activities have displaced the natural balance. Large amounts of methane are released by cattle farming, landfill waste dumps, rice farming, and the traditional production of oil and gas.

Nitrous oxide is produced through the large-scale use of commercial and organic fertilizers driving the microbial processes of nitrification and denitrification occurring in soils, ocean and freshwater systems, fossil-fuel combustion, nitric-acid production, and biomass burning. It is the third most important long-lived GHG after CO_2 and CH_4.

- **Carbon dioxide mapping and monitoring.** The Greenhouse Gases Observing Satellite (GOSAT) was the first satellite launched by Japan in 2009 to monitor atmospheric carbon dioxide (CO_2) and methane (CH_4). In 2014, the US National Aeronautics and Space Administration (NASA) launched a satellite, the Orbing Carbon Observatory 2 (OCO-2) dedicated to monitoring the atmospheric CO_2 and providing high-resolution observations. The PRC CO_2-monitoring satellite, TanSat, launched in December 2016, is also capable of providing space-based CO_2 measurements and preliminary XCO2 maps generated using satellite measurements. GOSAT-2 was successfully launched in October 2018 and OCO-3 in May 2019 to monitor atmospheric CO_2. Global CO_2 and CH_4 natural fluxes and anthropogenic emissions and their trends in support of the Paris Agreement are provided by the Copernicus Atmosphere Monitoring Service (CAMS).[155]

  The world's first commercial satellite dedicated to monitoring CO_2 from orbit was launched in November 2023 by the Canadian company GHGSat, which already flies six spacecraft tracking methane emissions. The new platform (GHGSat-C10) will use the same shortwave infrared sensor

[155] Copernicus Atmosphere Monitoring Service (CAMS).

but will be tuned to CO_2's specific light signature in the atmosphere.[156] The satellite will have a resolution of 25 m at ground level, meaning it will be the first satellite able to see major individual CO_2 sources like refineries, steel mills, aluminum smelters, cement plants, and thermal power stations.

The Copernicus Anthropogenic Carbon Dioxide Monitoring (CO2M) is a European Space Agency (ESA) mission that will be the first to measure how much carbon dioxide is released into the atmosphere, specifically through human activity.[157] Three satellites are planned, CO2M-A, CO2M-B, and CO2M-C, which will all be in sun-synchronous orbits at an altitude of 735 km and an orbital inclination of 97.7°. The orbital period of these satellites will be 99.5 minutes with an 11-day repeat cycle. The mission is a part of the Copernicus Sentinel Expansion mission, which ESA is developing on behalf of the European Union (EU) and consists of three satellites: CO2M-A, CO2M-B and CO2M-C, the first of which is planned for launch in 2025. The Integrated CO_2 and NO_2 Imaging Spectrometer (CO2I) will provide CO_2, CH_4, and NO_2 observation with a relatively high spatial resolution (4 km x 4 km) in support of estimating anthropogenic emissions.

Figure 24: The Carbon Cycle

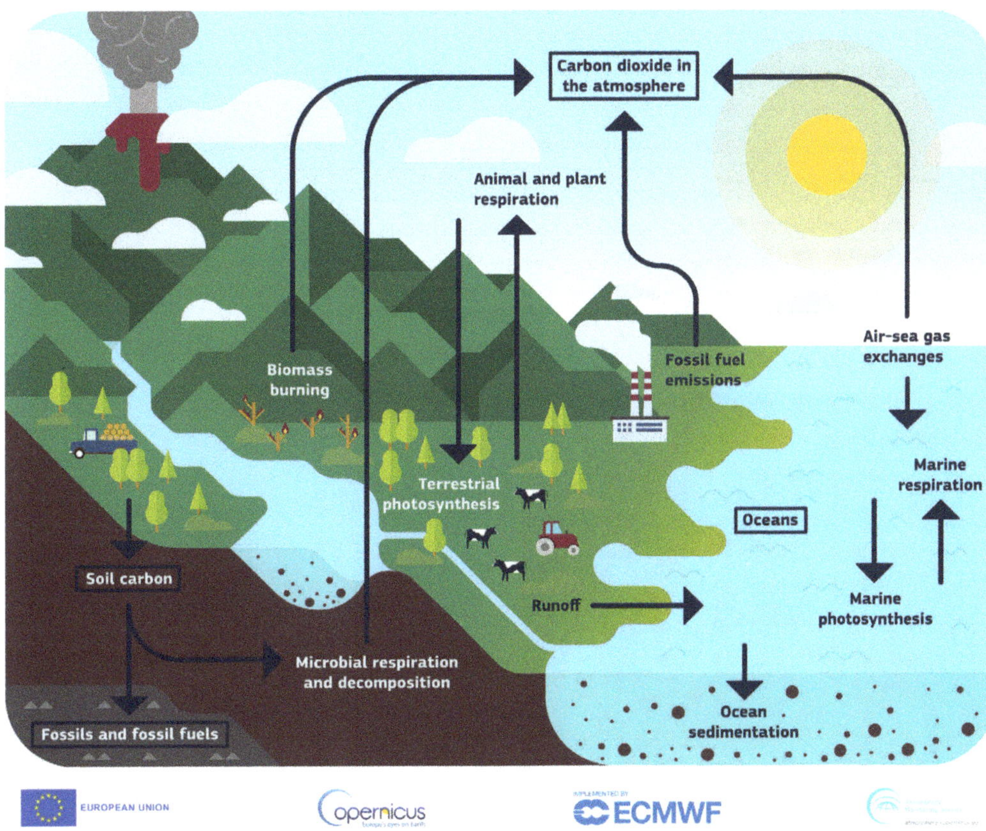

Source: The Copernicus Atmosphere Monitoring Service, ECMWF. https://atmosphere.copernicus.eu/ghg-services.

156 *Greenhouse Gas Emissions Monitoring Satellite* (GHGSat). 2023. GHGSat to launch world's first commercial CO2 satellite. 31 January.

157 eoPortal. CO2M (Carbon Dioxide Monitoring) Mission.

The figure depicts the growth in global atmospheric CO_2 over the last 2 decades.[158]

Figure 25: Satellite Data Record of Atmospheric Carbon Dioxide

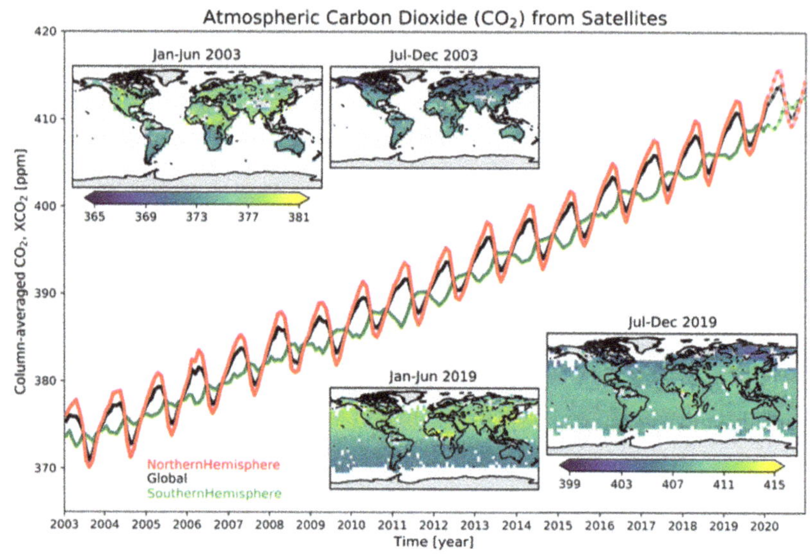

Sources: C3S, CCI, CAMS, University of Bremen, and SRON. https://atmosphere.copernicus.eu/cop26-cams-help-measure-progress-towards-co2-goals.

Map 66: Map of Monitoring Carbon Dioxide from Space

Sources: EMPA, ESA, CAMS, and ECMWF. https://atmosphere.copernicus.eu/ghg-services/observing-greenhouse-gases.

An example map of Global CO_2 distribution showing significant CO_2 concentrations over central and eastern India that are migrating eastward.[159]

[158] CAMS. Observations of greenhouse gases.

[159] CAMS. GHG Services.

Figure 26: Annual Mean Carbon Dioxide Exchange Between Land and Atmosphere for Agricultural, Forestry, and Other Land Use Sector

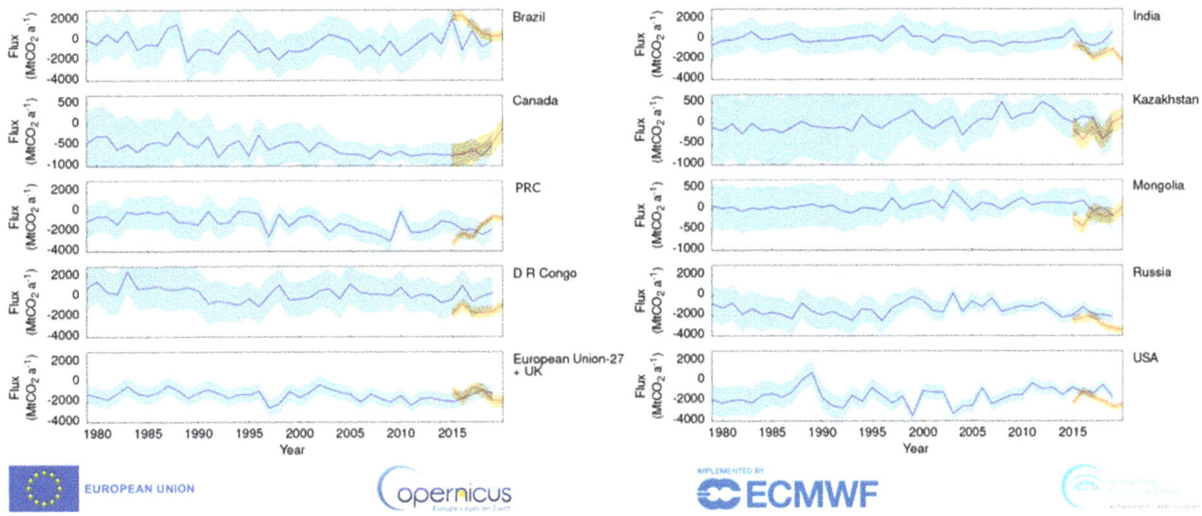

PRC = People's Republic of China.

Source: CAMS and CEA. https://atmosphere.copernicus.eu/cop26-cams-help-measure-progress-towards-co2-goals.

The annual mean CO_2 exchange between land and atmosphere for the agricultural, forestry, and other land use (AFOLU) sector in 10 large countries or groups of countries is estimated by CAMS using ground-based CO_2 observations (blue line) and satellite observations (orange line).[160] Positive values indicate that the AFOLU sector in that country is a source of CO_2 to the atmosphere. The uncertainty of the estimates is indicated by the colored areas.

ADB's GeodigitalTwin provides maps and analytics of three major CO_2-related key parameters: building emissions, transport emissions, and forest sequestration. Forest carbon stock and urbanized areas are estimated with satellites. CO_2 emissions are calculated via integration with Google Environmental Insight Explorer (a global dataset). From this dataset, the traveled distance and number of trips are used to calculate the amount of CO_2 emissions. Note that only for the Sumatra Islands, 1,600 urban settlements are monitored, giving a good spatial coverage.

[160] *CAMS.* 2021. COP26: CAMS to help measure progress towards CO_2 goals. 2 November.

Map 67: Estimating Carbon Dioxide Emissions in Sumatra

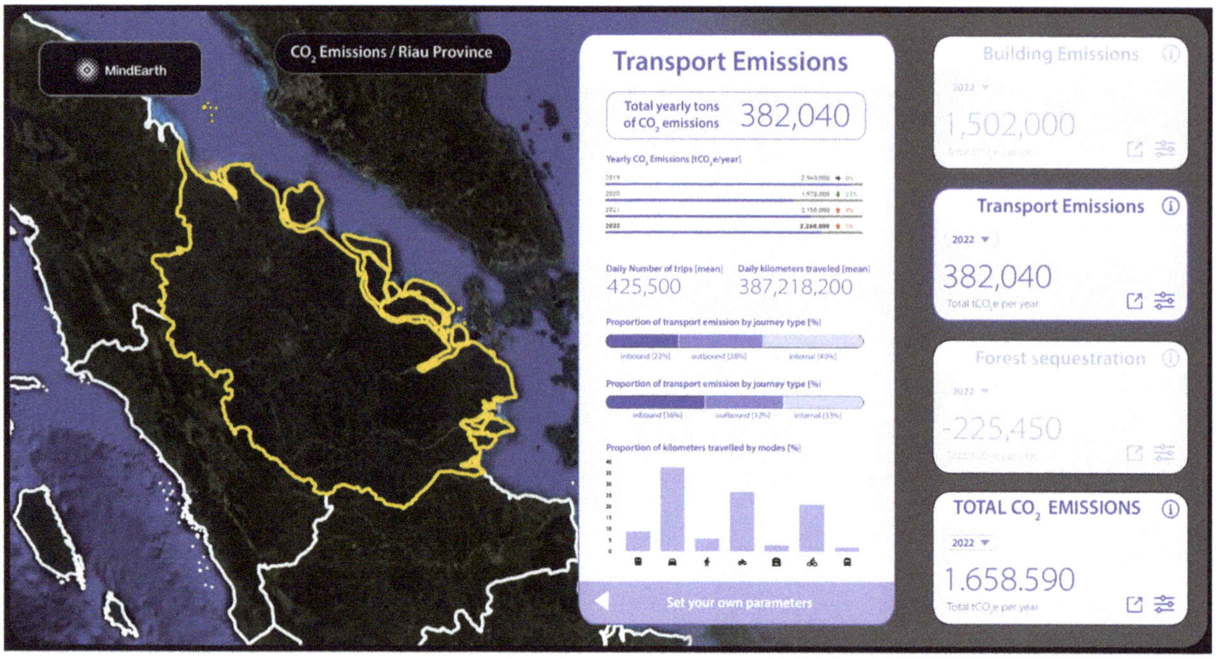

Sources: MindEarth, Switzerland; Asian Development Bank.

- **Methane mapping and monitoring.** Satellite radiance measurements in the Short Wave Infra-Red (SWIR) spectral region are sensitive to near-surface CH_4 concentration variations and therefore permit to retrieve column-average dry-air mole fractions of CH_4 (denoted XCH_4). Currently, the production and provision of these datasets are being generated (pre-) operationally using the ESA Envisat mission (historical Sciamachy instrument data), the Copernicus Sentinel-5P satellite, and the China National Space Administration's TanSat mission. The CH_4 climate datasets are provided via the Copernicus Climate Change Service (C3S), which is implemented by the European Centre for Medium-Range Weather Forecasts (ECMWF) on behalf of the European Commission.

Figure 27: The Methane Cycle

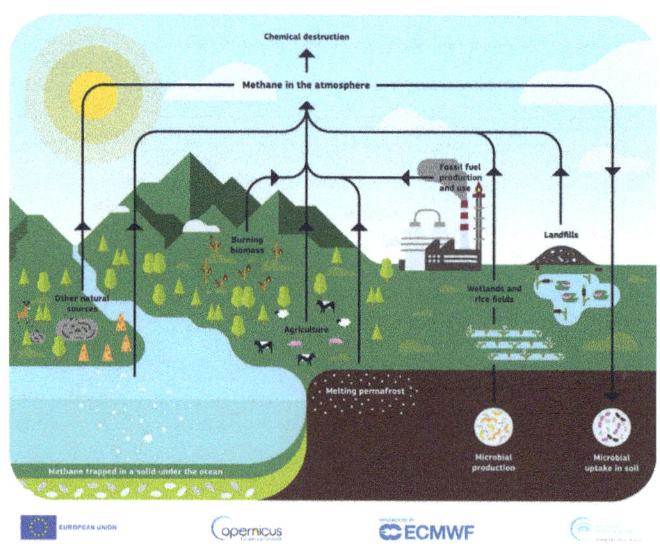

Source: Copernicus Atmosphere Monitoring Service, ECMWF. https://atmosphere.copernicus.eu/ghg-services.

Atmospheric methane has increased by about 150% since pre-industrial times, but concentrations have been nearly constant since the late 1990s. However, since 2007, CH_4 concentrations have started to increase again at a rate of about 8 parts per billion per year.

Figure 28: Satellite Data Record of Atmospheric Methane

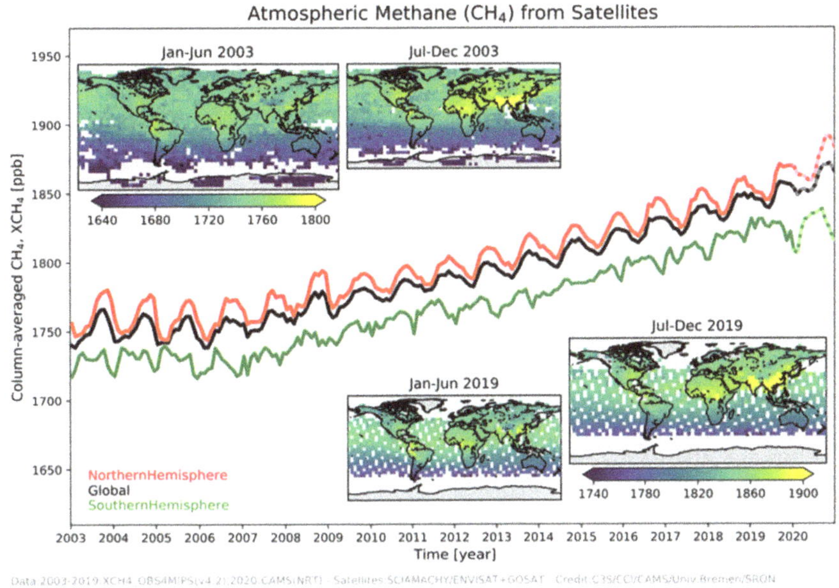

Source: C3S, CCI, CAMS, University of Bremen, SRON. https://atmosphere.copernicus.eu/ghg-services/observing-greenhouse-gases.

Map 68: Global Methane Forecast Total Column, 3 November 2022

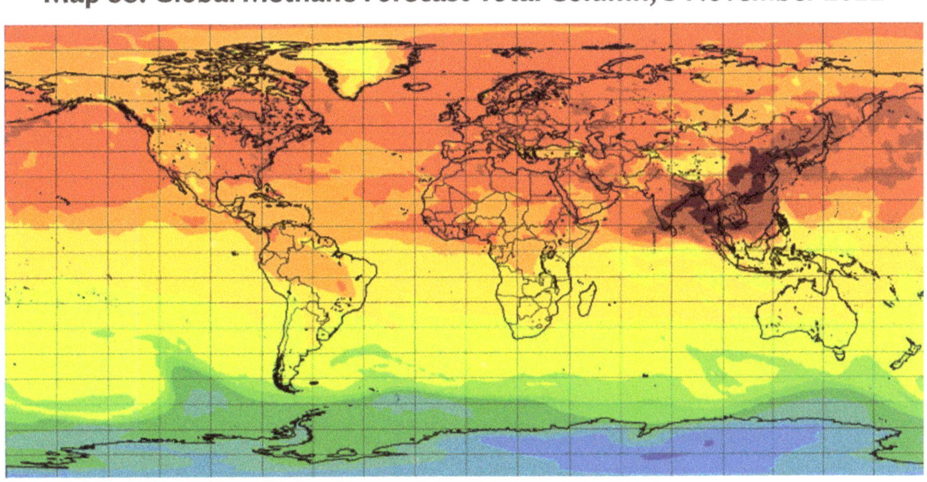

Source: Copernicus Atmospheric Monitoring Service (CAMs). https://atmosphere.copernicus.eu/methane-omnipresent-and-elusive-second-greenhouse-gas.

An example of a methane total column forecasting global distribution.[161]

[161] *CAMS*. 2022. Methane, the omnipresent and elusive "second" greenhouse gas. 4 November.

This map imagery was produced using the Copernicus Sentinel-5P Tropomi instrument. It produces a global map of methane concentrations every day.[162]

Super-emitters release a disproportionately large amount of methane compared to other emitters. Of the 2,974 plumes found in 2021, 45% originate from oil and gas facilities, 35% from urban areas, and 20% from coal mines.

Map 69: Global Overview of the Location and Magnitude of All 2,974 Methane Super-Emitter Plumes Detected in 2021

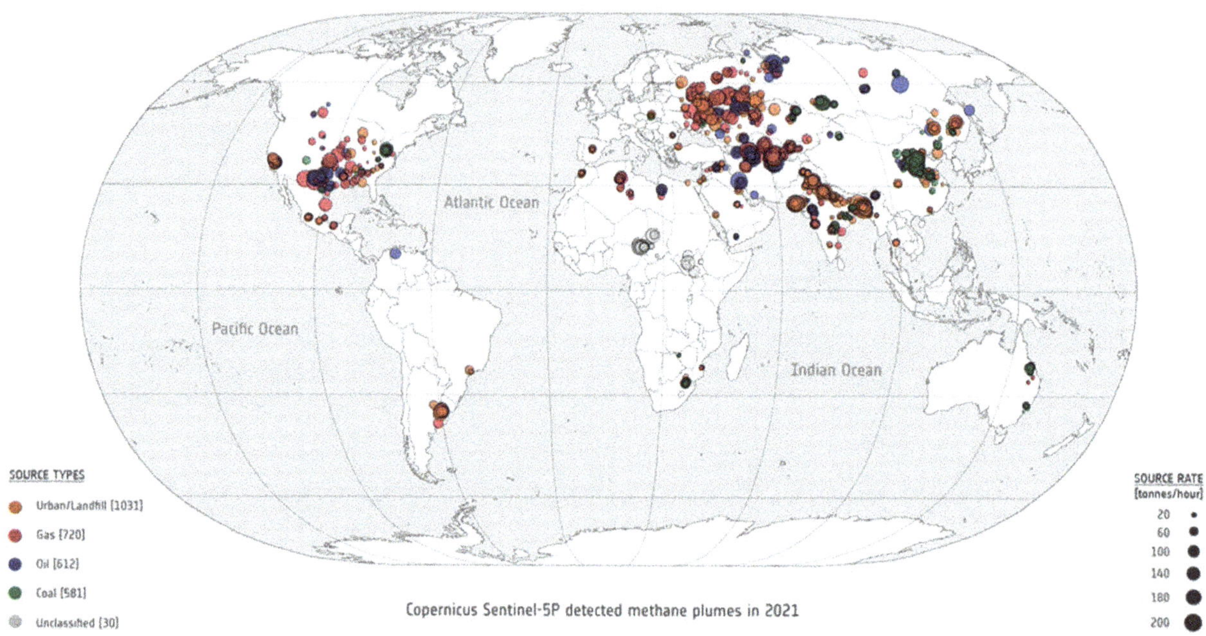

Source: ESA. https://www.esa.int/Applications/Observing_the_Earth/Copernicus/Trio_of_Sentinel_satellites_map_methane_super-emitters.

In addition to large space operators, a Canadian commercial satellite company, GHGSat, specializes in mapping and monitoring of GHG emissions globally via a constellation of SmallSats (GHGSAT-C1 to GHGSAT-C8, launched from 2020 to 2023).[163] Each satellite is equipped with a wide-angle Fabry-Perot (WAF-P) imaging spectrometer designed to measure the vertical column abundances of greenhouse gases at a high spatial resolution of 25 m and Field of View of approximately 12 km x 12 km. These datasets allow gas plumes emitted from industrial sources to be captured and distinguished from the surrounding background concentrations, constituting a differential measurement. The reflectance and abundance of products can be combined into a high readability "Concentration Map" for human interpretation.

162 ESA. 2023. Trio of Sentinel satellites map methane super-emitters. 20 September.

163 GHGSat.

The information shown in this map is from the GHGSat Methane Snapshot Report and was provided as input to COP27.[164]

Map 70: Geographical Distributions of Global Methane Emissions

Source: GHGSat. https://www.ghgsat.com/en/cop27/.

Oil and gas accounted for more than half of all emission observations in 2022 and 47% of the volume of emissions measured in 2022, up from 39% in 2021.

The information in this figure is from the GHGSat 2022 Methane Emissions Report.

Figure 29: Breakdown of 2022 Global Methane Emissions by Main Emitter Type

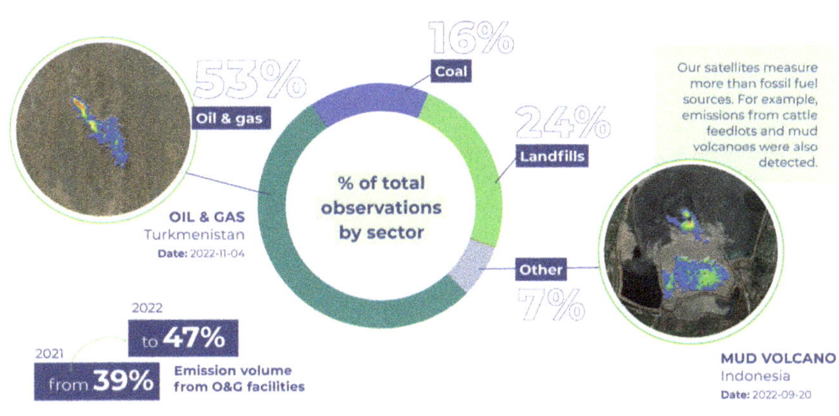

Source: GHGSat. https://www.ghgsat.com/en/case-studies/2022-methane-emissions-report/.

[164] GHGSat. COP27 – Global Methane Pledge.

Map 71: An Example of a Landfill Methane Emission from a Location in India

Source: GHGSat. https://www.ghgsat.com/en/case-studies/2022-methane-emissions-report/.

India has the most individual emissions and the largest estimated volume of CH_4 emissions coming from landfills, as detected by GHGSat satellites, according to the GHGSat 2022 Methane Emissions Report.[165]

GHGSat-C2 (Hugo), is the company's third satellite, and this is the first image of a methane plume from an oil and gas facility while overflying Asia in late January 2021, less than a week after launch.[166]

Map 72: Methane Emission from an Oil and Gas Facility in Asia

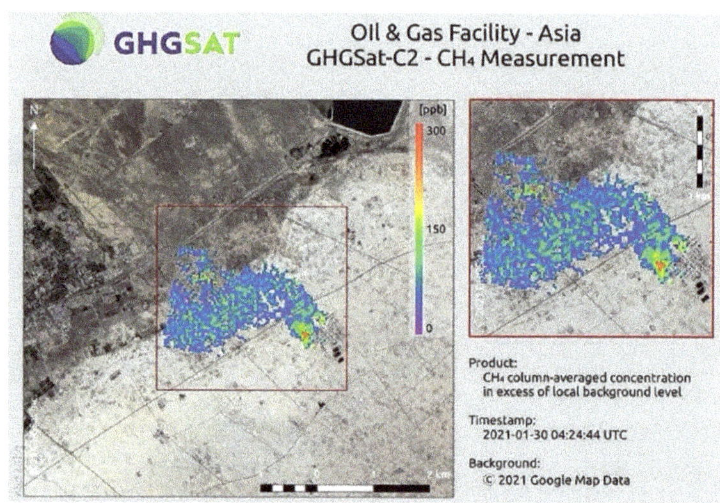

Source: GHGSat. https://www.eoportal.org/satellite-missions/ghgsat-con#mission-status.

165 GHGSat. 2022 Methane Emissions Report.

166 eoPortal. GHGSat Constellation.

- **Nitrous oxide mapping and monitoring.** Global satellite data can complement the sparse in-situ N_2O measurements, particularly in key regions that are remote and difficult to access, such as tropical areas. Since 2008, the Thermal InfraRed measurements from satellite instruments such as the Infrared Atmospheric Sounding Interferometer (IASI) on Metop-A, Atmospheric InfraRed Sounder (AIRS), and Greenhouse Gases Observing Satellite (GOSAT) have become available to observe the N_2O total column and upper tropospheric N_2O.

Map 73: Monitoring Nitrous Oxide Sources Globally at a Spatial Resolution of 1° × 1° and Regionally at a High Spatial Resolution of 10 × 10 Square Kilometers, July 2011

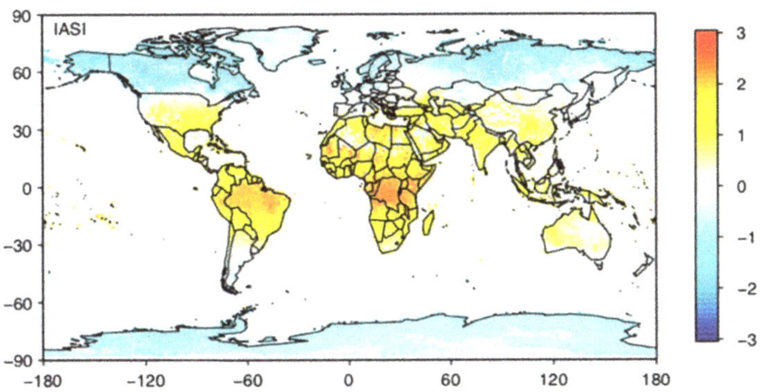

Source: P. Ricaud et al. 2021. The Monitoring Nitrous Oxide Sources (MIN2OS) satellite project. *Remote Sensing of Environment*. Vol. 266. 1 December.

This global map was produced by inverting surface fluxes from N_2O space-borne measurements (IASI on Metop-A) sensitive to the lowermost atmospheric layers.[167]

For anthropogenic emissions, East and South Asia, Europe, and North America are the most-emitting regions, while for natural soil emissions, Equatorial and South Africa and South America are the most-emitting regions. For the ocean, the Eastern Equatorial Pacific, South-Eastern Tropical Atlantic, and Eastern Indian Oceans are the main N_2O source areas.

The local higher N_2O columns in the tropics, especially over North Africa and offshore, are associated with the presence of atmospheric dust aerosols.[168]

Map 74: Global Maps of Retrieved Nitrous Oxide Partial Column from 800 to 80 hPa (monthly average) for (a) June and (b) December 2018

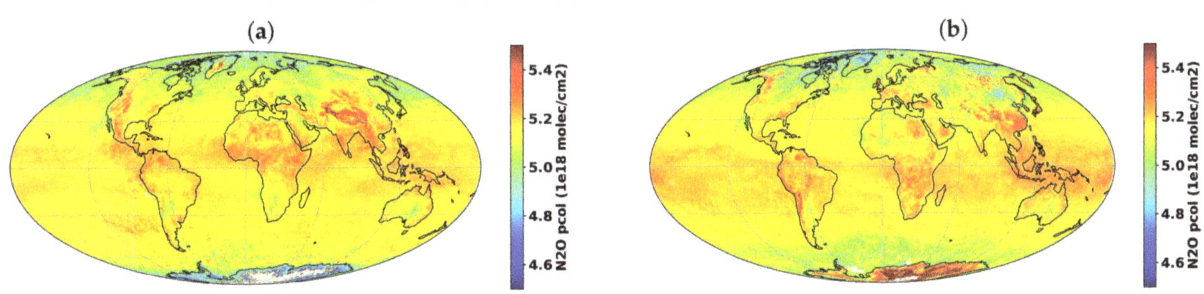

Sources: NOPIR Algorithm description; S. Vandenbussche et al. 2022. *Nitrous Oxide Profiling from Infrared Radiances (NOPIR): Algorithm Description, Application to 10 Years of IASI Observations and Quality Assessment (Figure 5)*. Remote Sensing. 14(8).

[167] ScienceDirect. 2021. *The Monitoring Nitrous Oxide Sources (MIN2OS) Satellite Project.*

[168] Multidisciplinary Digital Publishing Institute (MDPI). 2022. *Nitrous Oxide Profiling from Infrared Radiances (NOPIR): Algorithm Description, Application to 10 Years of IASI Observations and Quality Assessment.*

In summary, how are these types of GHG-related environmental information products generated from EO satellite data relevant to the development sector?

Satellite and ground-based observations are combined with atmospheric forecast models to support businesses, policymakers, and scientists dealing with the challenges and opportunities related to the composition of the atmosphere.

Understanding where GHG emissions come from is crucial in identifying actions to halt them and measure the impact of those actions to assess whether we are making the right changes that are quick enough.

Satellite GHG monitoring helps identify emissions from oil and gas sector, gas leaks, and coal mine; manage forestry; and better control waste management landfill emissions. This is being done with increasing frequency and spatial coverage.

As a practical application of how CO_2 emission information derived from satellite EO is being used in development, ADB has developed a digital analytics platform to help government officials simulate local policies that can generate different CO_2 scenarios toward carbon neutrality. An example of the types of information that this analytics platform can produce is given in Figure 30.

Figure 30: Example Output Information from an ADB Digital Platform for Carbon Dioxide Emissions Analytics

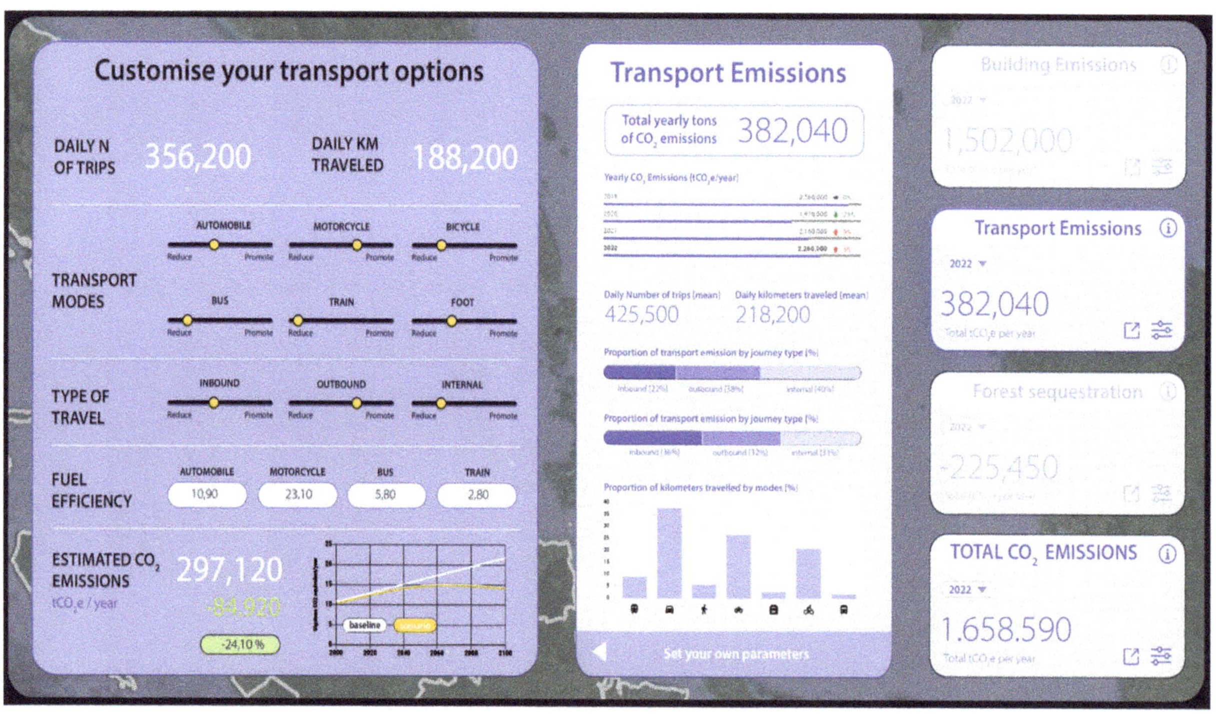

Source: Asian Development Bank, data processed by MindEarth, Switzerland.

Different transport options accompanied by investments can lead to lower emissions. Long-term development plans that affect land use can be also accounted for. This includes locations of major industrial sites, availability of work force nearby to minimize the travelled distance, or location-optimized supply and production chains.

The monitoring of greenhouse gas (GHG) emissions is directly connected to one of the seven ADB Strategy 2030 operational priorities: C. *Tackling Climate Change, Building Climate and Disaster Resilience, and Enhancing Environmental Sustainability*.

3.7 Earth Observation for Development Monitoring: A Use Case on the Urban Environment

EO can provide key environmental information to help in monitoring the UN Agenda 2030 for Sustainable Development (Agenda 2030), which consists of 17 high-level Sustainable Development Goals (SDGs), with 169 medium-level associated targets further broken down and supported by 232 detailed-level indicators.

The overall contribution and organizational framework of how EO is contributing to Agenda 2030 has been described in an ADB working paper, *Satellite Earth Observation Assets and Solutions*.[169]

This section presents a more detailed example of how a specific EO-based environmental information can help monitor and understand the changes taking place that are associated with a specific SDG target and indicator.

11 SUSTAINABLE CITIES AND COMMUNITIES

TARGET 11.1

INDICATOR 11.1.1: PROPORTION OF URBAN POPULATION LIVING IN SLUMS, INFORMAL SETTLEMENTS, OR INADEQUATE HOUSING

By 2030, ensure access for all to adequate, safe and affordable housing and basic services and upgrade slums

Informal Settlements and Urban Growth Patterns in Metro Manila, Philippines

The Philippines is one of the fastest urbanizing countries in Southeast Asia, with its capital Metro Manila, accounting for 30% of the country's population and 50% of economic output, while occupying less than 4% of the national territory.

Metro Manila faces many urban challenges, including poverty, informal settlements, and significant risks from natural hazard events (mainly flooding due to typhoons and tropical storms).

169 P. Manunta and S. Coulson. 2024. Satellite Earth Observation Assets and Solutions: Supporting Operations in Asia and the Pacific. *ADB Sustainable Development Working Paper Series*. No. 99.

Map 75: Area of Interest, Metro Manila, Philippines

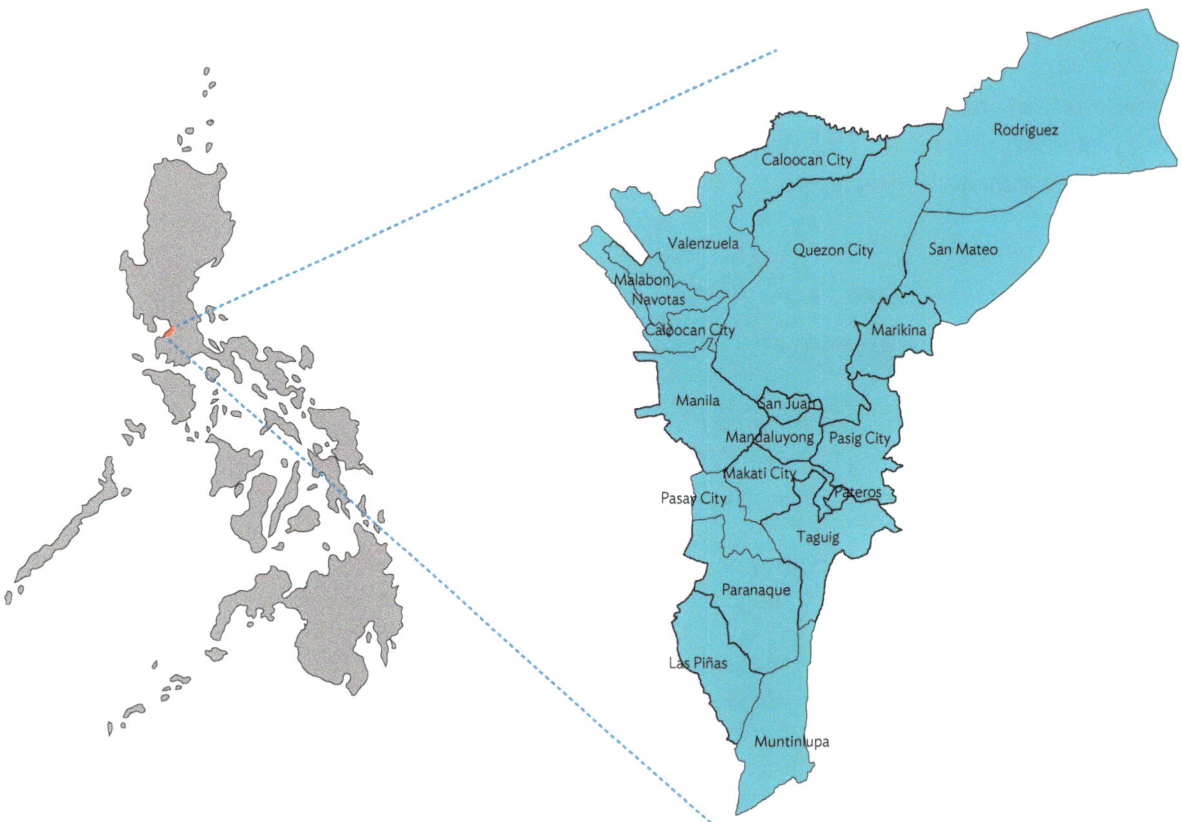

Source: European Space Agency (ESA).

In 2014, the European Space Agency (ESA) commissioned a study to assess how EO information could help in dealing with these specific urban challenges and therefore lead and support the government to progress with Metro Manila's long-term 2030 vision for city development.[170]

The study was carried out by GIM, a small specialist geo-ICT company located in Belgium, and covered the entire Metro Manila area plus parts of Rizal province (see above maps). The main types of urban information produced were

- land use and land cover (LULC) mapping for four points in time (1990–2000 to 2010–2014);
- change detection and analysis on the LULC time series to understand urban dynamics; and
- informal settlement families mapping and characterization of their relationship to hazardous zones, and educational and health infrastructure.

[170] ESA and World Bank. 2016. *Earth Observation for Sustainable Development.* pp. 78–83.

These types of urban information require the use of multispectral satellite imagery. The main sources of data used was from the French national missions of high-resolution (HR, 20 m) SPOT-1, SPOT-2, SPOT-4, and very-high-resolution (VHR, 0.5 m) Pléiades-1A satellites.

The maps below show the basic land surface classification mapping of Metro Manila. Land use classes and the evolution in time of the land area coverage statistics for specific land use classes are also defined.

Map 76: Land Use and Land Cover Maps and Land Use Classes Evolution Statistics (1990, 2000, 2010, and 2014)

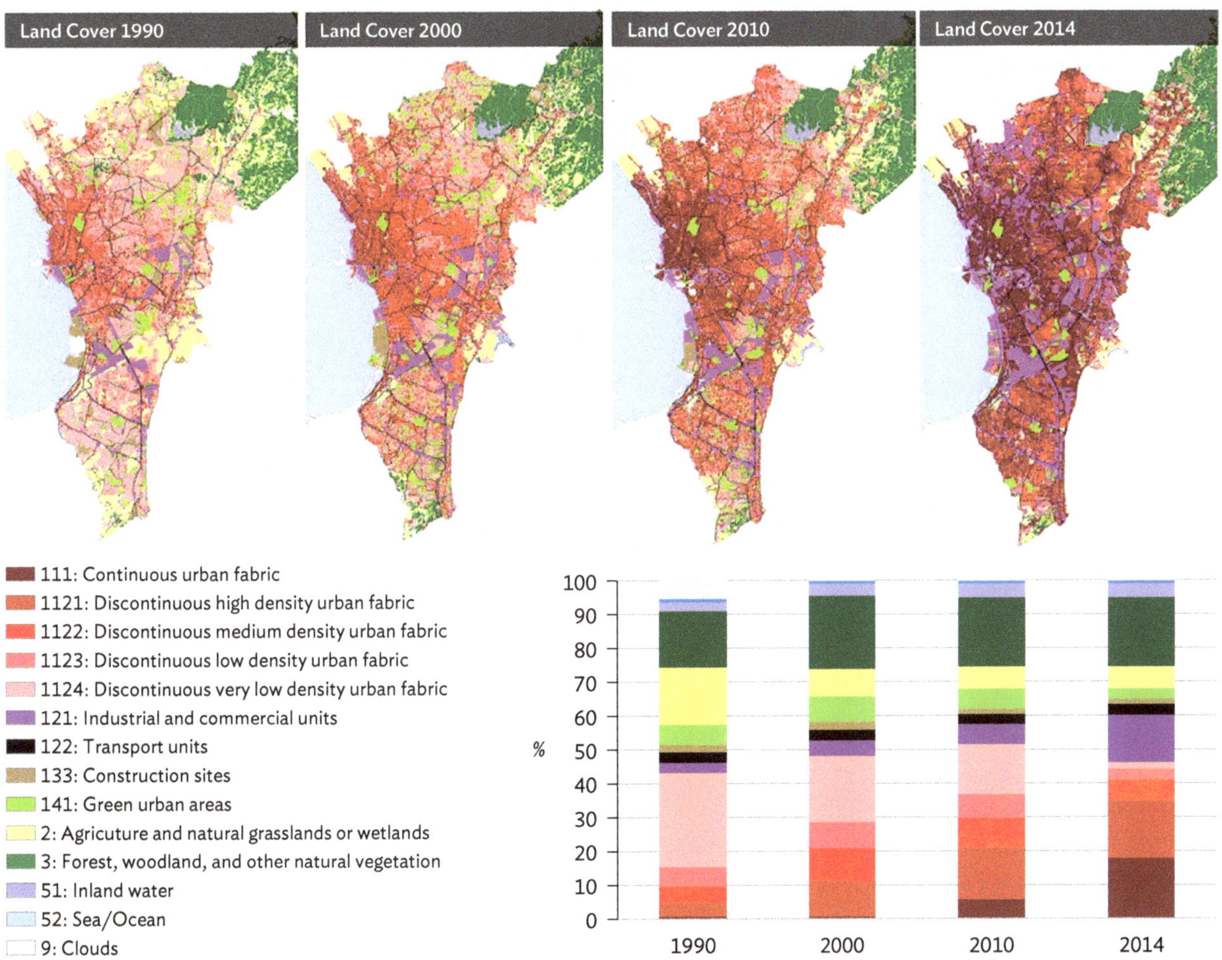

Sources: GIM and ESA. Note that the satellite imagery is from SPOT 1, 2, and 4 HRV series, © CNES 2014. European Space Agency. 2016. *Earth Observation for Sustainable Development – Partnership Report*. p. 28.

These basic land surface classification products can then be used for subsequent spatial analyses such as the detection and analysis of changes occurring throughout the period, and the identification of specific urban growth patterns and densification trends.

This analysis can be carried out for the entire Metro Manila area, or concentrated on specific locations (see Figure 31 which focuses on Caloocan City).

Figure 31: Land Use Land Cover Distribution of Aggregated Classes in Caloocan City

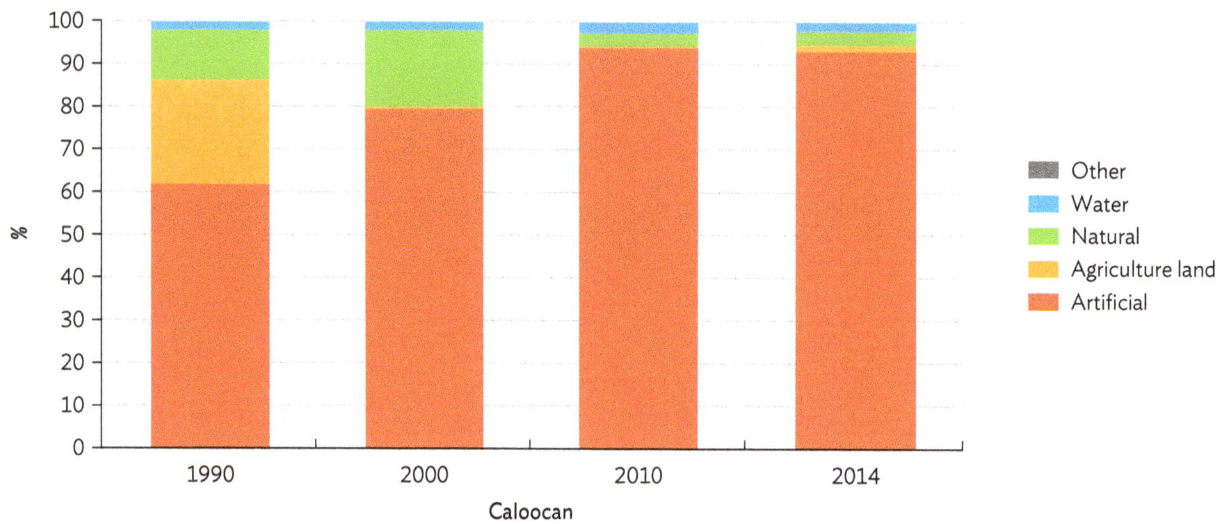

Sources: GIM and ESA. https://eo4society.esa.int/wp-content/uploads/2021/11/EOSD_160905-V2-FINAL.pdf.

For the period 1990–2000, the expansion of the city was significant at 17.7% and the urban growth continued at 14.3% for the period 2000–2010.

Cities showing the highest urban sprawl (more than 10%) over the years are Caloocan City, Pasig City, Taguig City, and Valenzuela City. The highest urban sprawl for these cities occurred in 1990–2000. Urban sprawl is a result of the conversion of any non-artificial LULC class to artificial classes with urban green not being considered as part of artificial classes.

To address the lack of detailed, consistent, comprehensive, and up-to-date information on informal settlement families in Metro Manila, VHR multispectral imagery (Pleiades 0.5-meter data for 2014) was processed to detect and delineate all informal settlements in a semiautomated method over the entire area of interest.

Informal settlements possess specific physical and structural characteristics that distinguish them from other residential areas (i.e. the extent and form of housing structures, density, size and number of streets, absence of regular spatial patterns, irregular boundaries). These characteristics can be extracted and measured from remote sensing data. However, due to the complexity associated with identifying an informal settlement family, these results need to be checked and validated using in-situ sampling data.

The total number of informal settlements detected in the area of interest is 2,443 with a total area of 29.281 square kilometers, which represents 3.77% of the total area of interest. However, the informal settlements are not distributed evenly among the municipalities of Metro Manila.

Map 77: Geographic Location and Types of Informal Settlement Families in Metro Manila, 2014

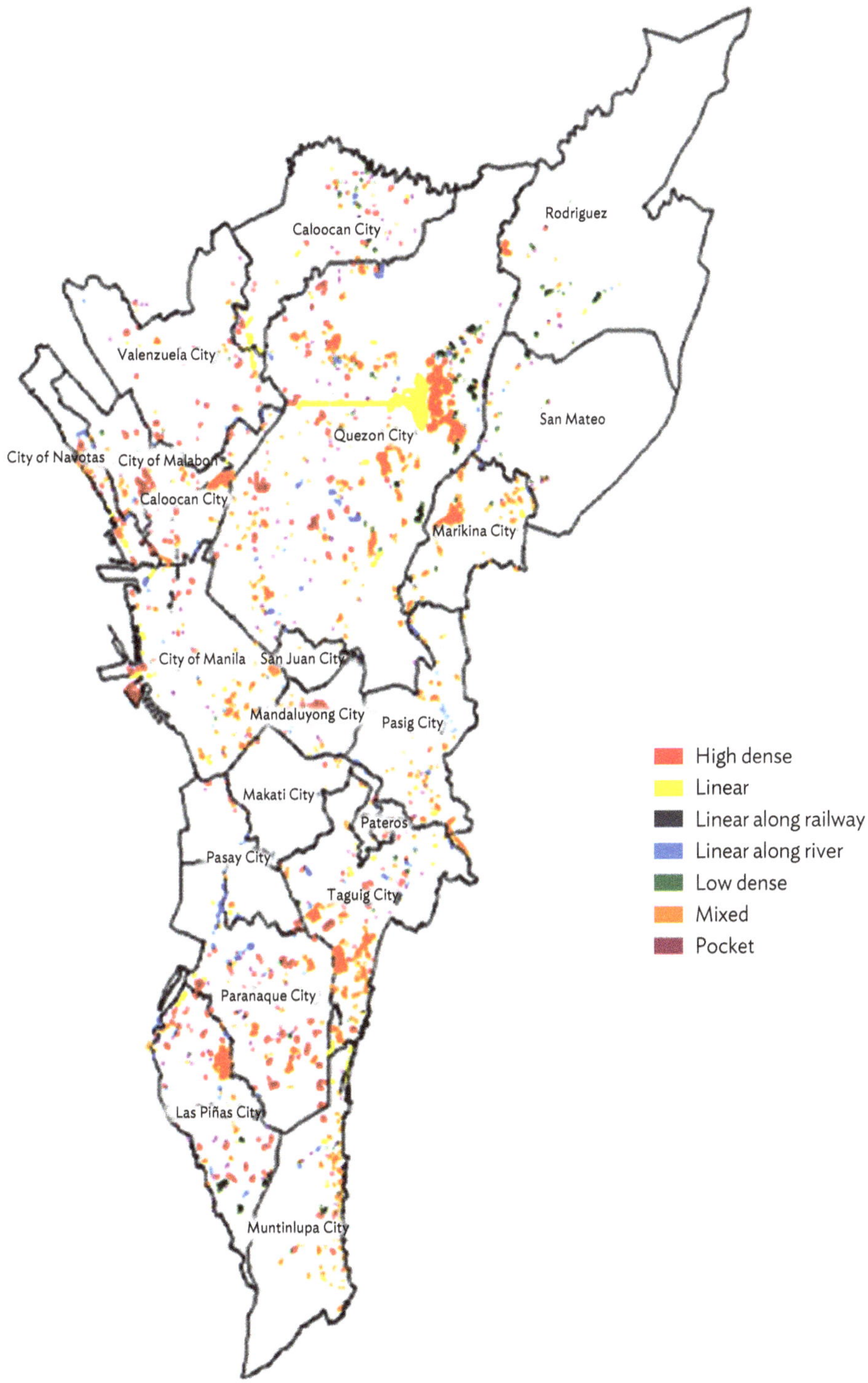

Sources: GIM and ESA. https://eo4society.esa.int/wp-content/uploads/2021/11/EOSD_160905-V2-FINAL.pdf.

Figure 32: Informal Settlement Families Distribution per Slum Type for Metro Manila, 2014
(%)

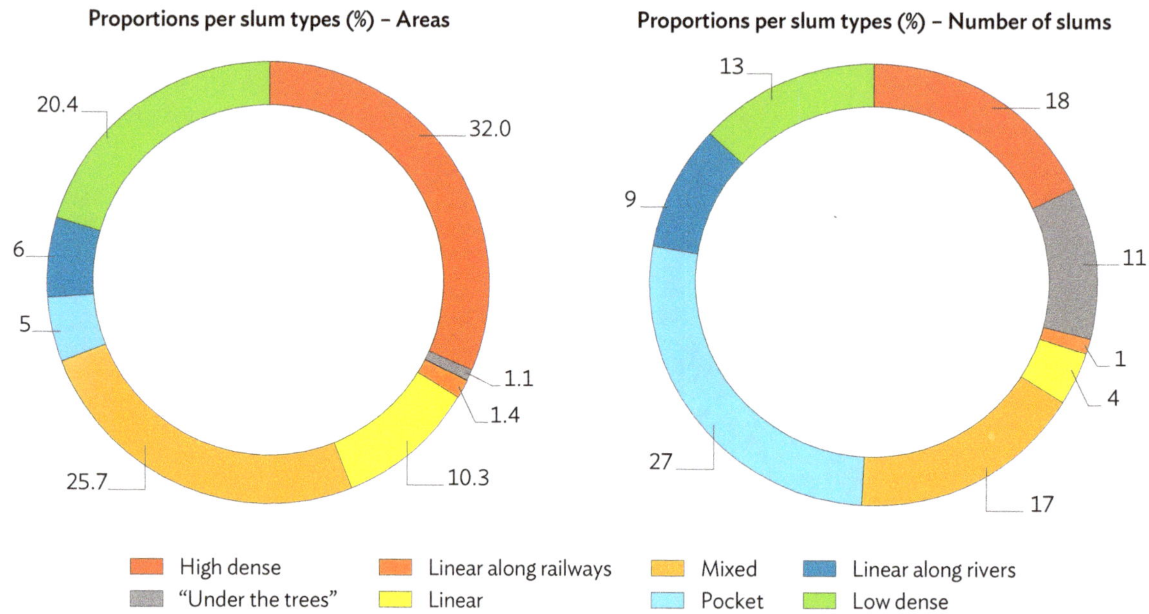

Sources: GIM and ESA. https://eo4society.esa.int/wp-content/uploads/2021/11/EOSD_160905-V2-FINAL.pdf.

Calculating the ratio of area of informal settlement family (ISF) per city to area of ISF in the total area of interest shows that Quezon City has the biggest proportion of ISF (33.89%, but is also the largest municipality). This is followed by Taguig City which includes 10.06% of the total area of the ISF. San Juan City, Makati City, and Pateros have less than 1% of the total area of the ISF in the area of interest.

Further spatial analysis was carried out to combine the ISF geographic location information with three additional existing types of local information:

- results of the hydrological simulation of the flood extent for an event characterized by a return period of 50 years, which allows assessment of water depth in different parts of flooded settlements and what percentage of informal settlements would be affected;
- location of schools, which allows assessment of access to education of people living in informal settlements; and
- location of hospitals, which allows assessment of access to health care of people living in informal settlements.

For the results of these analyses, see Figures 33 to 35.

Figure 33: Proportion of Informal Settlement Family Areas That Are Flood Prone

(%)

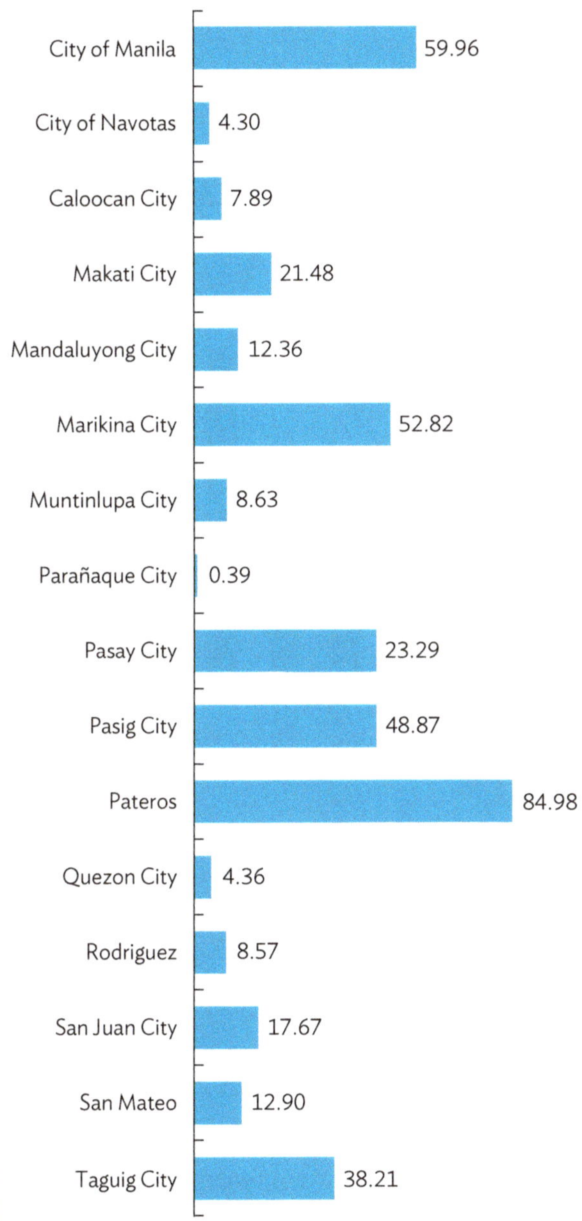

Sources: GIM, ESA (Note: For an event with a return period of 50 years; only the municipalities for which data was available were considered).

Figure 34: Proportion of Informal Settlement Within 2-Kilometer Distance from a Large Hospital

(%)

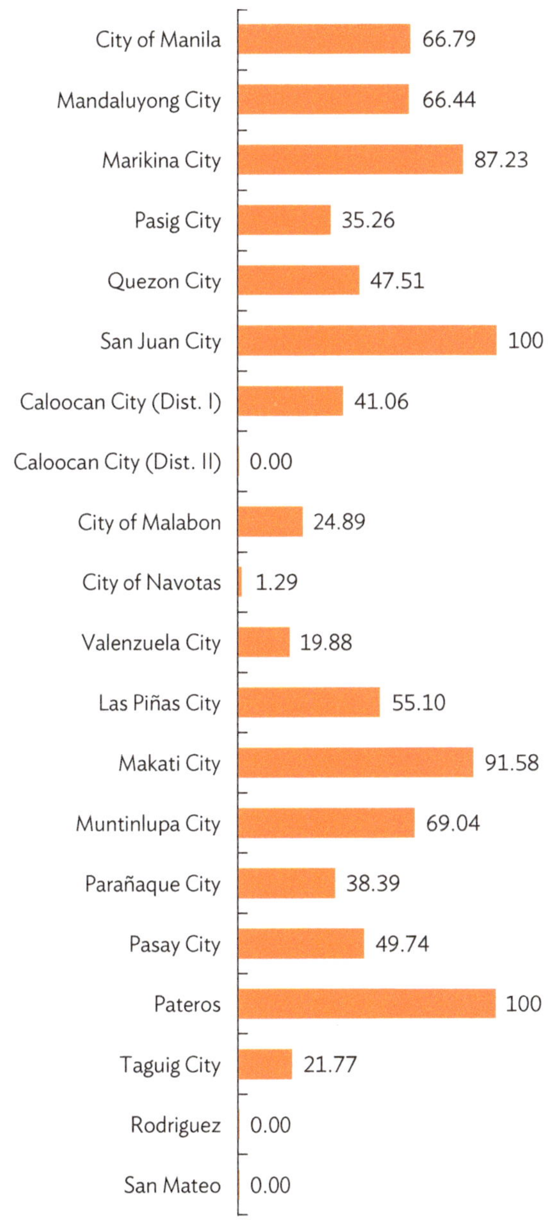

Sources: GIM, ESA.

In total, 38.07% of ISFs in Metro Manila are found within 1 km from a school, and 43.9% are found within 2 km from a large hospital.

However, the location of the hospitals comes from Open Street Map, and its accuracy of information and degree of completeness is not known; therefore, this analysis is made only for demonstration purposes and firm conclusions should not be drawn from it.

Figure 35: Proportion of Informal Settlements Within 1-Kilometer Distance from a School

(%)

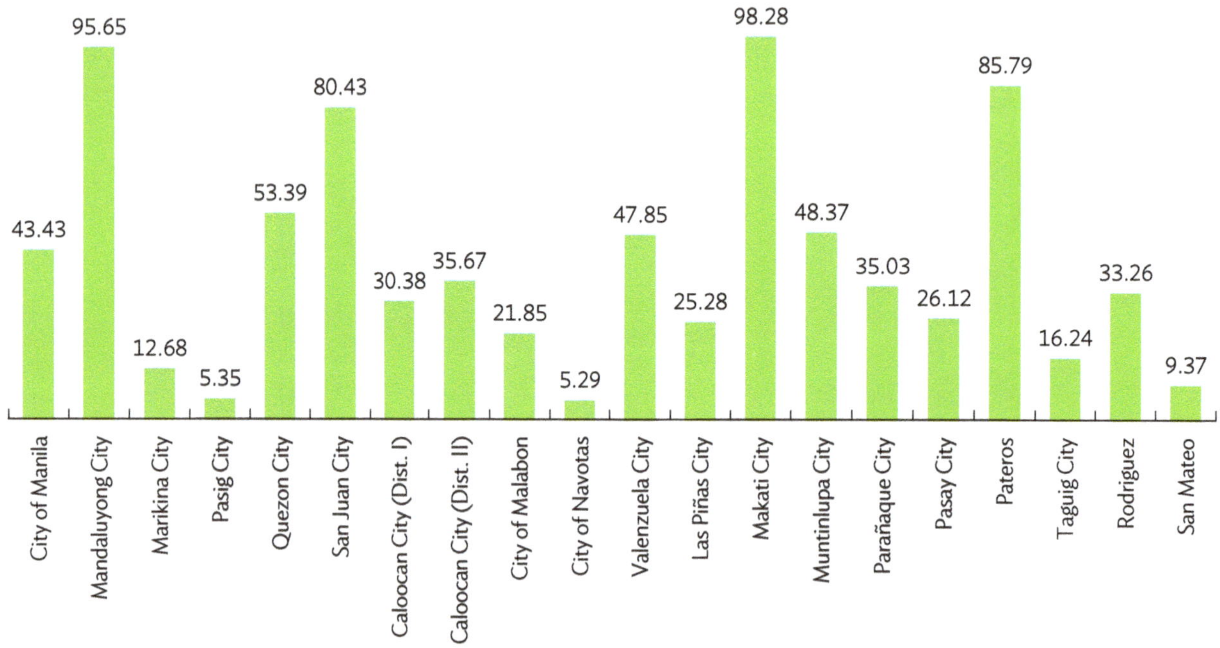

Sources: GIM, ESA.

In summary, EO and geospatial information can significantly contribute to monitoring and reporting of SDGs and can reduce the costs of this monitoring, making them more viable within the limited resources available to governments.

CONCLUSIONS

4

The main conclusions arising from the materials presented in this technical study are as follows:

(1) The most comprehensive, complete, and capable Earth Observation (EO) satellite system in the world today is Copernicus, the EO flagship program of the European Union (EU). Copernicus is a long-term program that started in 2014 to develop, build, and launch a constellation consisting of almost 30 Sentinel satellites in orbit before 2030. The Sentinel satellite families monitor the planet globally, consistently, and with frequent revisit. The Copernicus system provides free EO data and thematic information services based on satellite EO data and in-situ data (i.e., non-space).

(2) High-income economies have made substantial investments in satellite EO over the last 50 years. These capabilities contribute to global efforts in environmental monitoring, climate variability assessment, disaster management, and sustainable resource utilization.

(3) Developing economies in Asia and the Pacific have made significant strides in satellite EO with advanced capabilities in the development, production, and operation of microsatellites and cube satellites (CubeSats). They are harnessing space-based technologies for a wide range of applications, including disaster management, agriculture, natural resources monitoring, environmental protection, and climate resilience through dedicated ministries and departments. This represents an excellent foundation to consolidate and expand regional collaboration and cooperation to further grow the use of EO in the region.

(4) However, over the last decade, the satellite EO domain has transformed through a range of exciting, disruptive, and highly innovative development known as NewSpace. This emerging model of NewSpace is characterized by the proliferation of private companies and small startups entering the space industry with radically different approaches, technologies, and business models.

(5) NewSpace is driven by constellations of many micro and nano satellites, built mainly with commercial components and through simplified production and test processes that rely on low complexity. This has led to cheaper manufacturing costs and faster timescales to put satellite constellations into operation, which is attracting a wide range of private sector entrepreneurs focused on commercial objectives, and venture capital investments.

(6) This new trend began and is well advanced in the United States (US), with Europe now catching up. However, Asia and the Pacific is now poised to be the fastest-growing region in this innovative NewSpace technology sector over the next decade.

(7) As-a-service (AaS) business models are now being adopted across the EO value chain, including data analytics, payload, and ground segment. As such, the satellite data is usually not free and is available only at commercial rates, but data costs are rapidly falling.

(8) With the advent of high-performance cloud computing infrastructures and artificial intelligence (AI) data analysis techniques, complex environmental monitoring problems are beginning to be solved with EO-based information. New and innovative geospatial data processing infrastructures and platforms (e.g., data cubes, digital twins) are key to realizing the potential of EO, especially given the volume and complexity of EO data now being produced.

(9) EO value-adding and analytics environmental information services are generated and provided by a vast and highly diverse range of both public sector organizations and rapidly growing private sector companies.

(10) The EO data market currently stands at $1.78 billion (as of 2022), growing at a 5-year compound annual growth rate (CAGR) of 6%. Further future growth is expected and estimated to reach a total value of more than $2.7 billion by 2032, growing at a 4% CAGR throughout the decade, mostly driven by the demand for very high resolution (submetric) imagery.

(11) In parallel, the EO services and analytics market is currently valued at $2.86 billion (as of 2022) and demonstrates a 5-year CAGR of 9%. Predictions forecast an increase in total value to $4.9 billion by 2032, growing at a 6% CAGR.

(12) North America is the largest geographic region in terms of EO market share; however, Asia and the Pacific is estimated to grow, with the highest CAGR over the period 2023–2027.

(13) In 2019, an estimated 241 companies were active in the EO industry in developing economies, representing a total of €82 million, or 2.2% of the global EO market. This relatively low contribution, however, opens a significant opportunity for future growth in the digital economy of these developing economies.

(14) As the global economy continues to navigate the challenges posed by inflation, geopolitical tensions, and the residual effects of pandemic-related disruptions, the EO sector is undergoing a rapidly transformative phase, with commercial operators restructuring, reducing costs, and optimizing resources, with a strong focus on execution in an increasingly competitive landscape.

(15) Across the broader geospatial industry, the next 5 to 10 years will see significant developments in the maturity and application of already well-established technologies. Among others, AI, sensor technology, and the Internet of Things will drastically change the way in which data are collected, managed, and maintained.

(16) Satellite EO provides a wide range of key thematic types of environmental information that is global, comprehensive, accurate, repeatable, and timely. Satellite EO remains the only means to provide a comprehensive, consistent, and up-to-date imagery of the Earth's global surface and of the changes taking place on that surface both in time and space.

(17) Over the last decade, ADB has been exploring the potential of environmental information derived from satellite EO image data to be used in development operations, planning, and policymaking.

(18) Regarding the seven operational priorities of ADB Strategy 2030, EO is already making important and useful contributions to several of these (as identified in the ADB use cases) and has the potential to grow significantly in the future.

(19) EO and geospatial information can significantly contribute to monitoring and reporting of UN Agenda 2030 Sustainable Development Goals and associated targets and indicators, and can reduce the costs of this monitoring, making them more viable within the limited resources available to governments.

www.ingramcontent.com/pod-product-compliance
Lightning Source LLC
Chambersburg PA
CBHW041755220926
51847CB00014B/210
* 9 7 8 9 2 9 2 7 7 7 3 0 2 *